EXPERIMENTS AND DEMONSTRATIONS IN PHYSIOLOGY

Stephen E. DiCarlo
Eilynn Sipe
J. Paul Layshock
Rebecca L. Rosian

PRENTICE HALL Upper Saddle River, NJ 07458

Acquisition Editor: *Linda Schreiber*
Production Editor: *Dawn Blayer*
Manufacturing Buyer: *Ben Smith*
Special Projects Manager: *Barbara A. Murray*
Supplement Cover Manager: *Paul Gourhan*
Supplement Cover Designer: *PM Workshop Inc.*

A Pearson Education Company
Upper Saddle River, New Jersey 07458

Printed in the United States of America

ISBN 0-13-636457-8

Prentice-Hall International (UK) Limited, *London*
Prentice-Hall of Australia Pty. Limited, *Sydney*
Prentice-Hall Canada, Inc., *Toronto*
Prentice-Hall Hispanoamericana, S.A., *Mexico*
Prentice-Hall of India Private Limited, *New Delhi*
Prentice-Hall of Japan, Inc., *Tokyo*
Simon & Schuster Asia Pte. Ltd., *Singapore*
Editora Prentice-Hall do Brasil, Ltda., *Rio de Janeiro*

CONTENTS

INTRODUCTION

Research has shown that students learn best when they become actively involved with their subject matter. In addition, when they can see a relationship between the topics and their personal lives lessons become much more interesting and lasting for students. Teaching a laboratory-based physiology course allows an instructor to satisfy both of these conditions. The material is relevant and interesting and provides an opportunity for hands-on activities.

Finding good physiology laboratory books can be especially difficult, because many of the lessons require equipment that most schools cannot afford. In addition, most of the exercises provide very few opportunities for individual investigation. It is with these points in mind that this series of laboratory exercises was designed.

The activities that make up this book are designed to allow the student to experience a variety of topics within the field of physiology and to develop essential skills used by scientists when conducting investigations. These skills include observing phenomena, measuring objects, organizing information, recording and graphing data, and analyzing results. In many of the activities, the students are given the chance to conduct their own investigation by choosing from a number of options. All the equipment needed to conduct the activities is limited to common items (like string, rulers, and straws) found in every home.

It is hoped that this project will give the student insight into how investigations are conducted as well as provide practice in thinking and analyzing. In addition, our goal was to introduce students to the joys, excitement, and mystery of physiology and to stimulate their interest in future study. Along the way, the students are introduced to basic physiological concepts and how their bodies work. The instructor and his or her class now have supplementary material that will leave their students confident, motivated, and excited about learning. The exercises are inexpensive, interesting, and enlightening.

LABORATORY EXERCISE 1

INCORPORATING STATISTICS INTO PHYSIOLOGY EXPERIMENTS

BACKGROUND AND KEY TERMS

Reliability is a measure of accuracy, dependability, and consistency. For example, a reliable person can be depended on to get a job done and to do it well. In much the same way, a reliable experiment or measuring device repeatedly gives the same results over time. Reliability needs to be incorporated into all procedures of an investigation to assure that the results obtained during the initial experiment can be accurately reproduced during succeeding experimental trials. For example, in Part I, one student will make measurements of another student's leg. The technique used to obtain the measurement must be consistent so that different students making the same measurements of the student's leg will obtain the same values.

The **mean** is defined as the average value of a set of numbers. If you were to average the scores from an exam taken by 20 students, you would first add together all the scores and then divide the sum by the number of scores (in this case the number of scores is also the number of students who took the exam). The mean value simplifies a data set so that one value represents a given population. For example, if the mean is found to be 85%, then most values are near 85%; however, some values are much higher (100%) or much lower (65%) than the mean.

The **median** is a number located exactly in the middle of a set of numbers when they are ranked in ascending (or descending) order. If your grade is above the median, then you know that you are in the top 50% of your class. The median gives you an idea of where a data point fits into the overall picture.

The **mode** is the value that occurs with the greatest frequency. It represents the most common response and indicates the popularity of a reply. For example, it is helpful for a car manufacturer to determine the mode in a survey of people that asks for their favorite car color.

The following table gives the results of a 5-point quiz given to 13 students:

QUIZ SCORE	FREQUENCY (Number of Students)
5	5
4	1
3	2
2	1
1	2
0	2

The results could also be presented as a list of all the scores ranked from highest to lowest (or lowest to highest); for example, 5, 5, 5, 5, 5, 4, 3, 3, 2, 1, 1, 0, 0.

The *median* (middle score) of the 13 quiz scores is 3. If there are an even number of scores, the median is the mean of the two middle scores.

The *mode* (most common score) in this example is 5. More students obtained a score of 5 than any other score.

The *mean* (average score) in this example, is 3:(5+5+5+5+5+4+3+3+2+1+1+0+0 = 39) divided by the number of scores (13).

The data obtained from laboratory experiments can be displayed graphically. Histograms (bar graphs) and line graphs are most often used. A histogram of the quiz results would appear as follows:

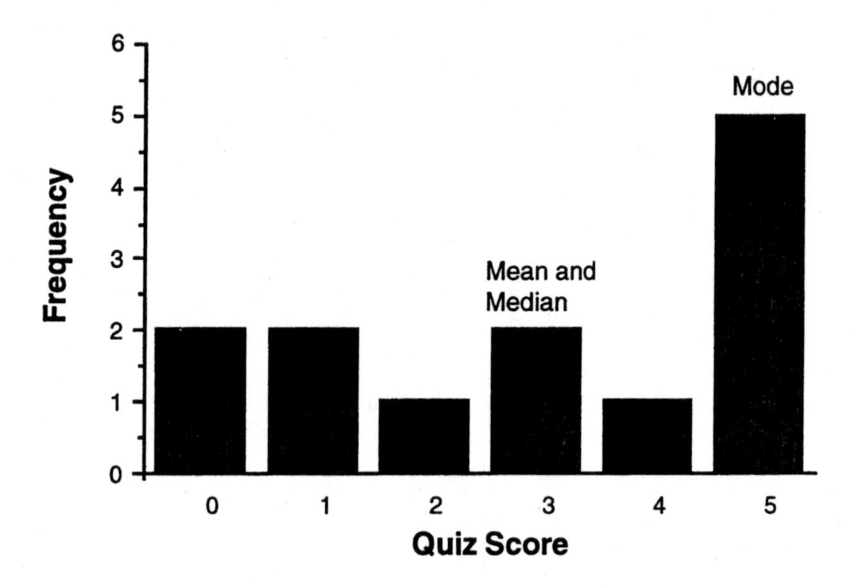

Scatter plots are diagrams in which points representing two measurements per subject are plotted on a pair of axes. Scatter plots are a good way to show the relationship between two variables. When all the collected data are plotted, a pattern may be found. For example, in Figure 2, if the points plotted on a graph resemble a straight line that is rising, a positive correlation exists between the two measurements. If the points resemble a straight line that is falling, then a negative correlation exists. If the points do not seem to follow any pattern, then zero correlation exists.

Figure 2. Positive, negative, and zero correlation

A **regression line** is a straight line drawn through the points in a scatter plot such that an equal number of points lie above and below the regression line (Figure 3a). It is also called the *line of best fit*. The importance of a regression line is that it is a tool that we can use to predict the values of one measurement when we are given the values of the other. For example, we can predict the number of push-ups a student can perform based on the number of sit-ups he or she can complete. For half the class we construct a scatter diagram in which each point represents the number of sit-ups and push-ups completed by one student. We draw a regression line through the points so that we can use the data to predict the number of push-ups the students in the rest of the class can perform. If an individual performs as many sit-ups as possible in 1 minute, we can predict the number of push-ups he or she can probably do from the regression line. We find the point on the regression line that corresponds to the number of sit-ups completed and read off the other axis the corresponding number of push-ups. Follow the arrow in Figure 3b to see that based on 45 sit-ups completed in 1 minute we would predict that 21–22 push-ups could be completed.

Figure 3a. Regression lines drawn through scatter diagram

Figure 3b. Predicting the number of push-ups based on number of sit-ups completed in 1 minute.

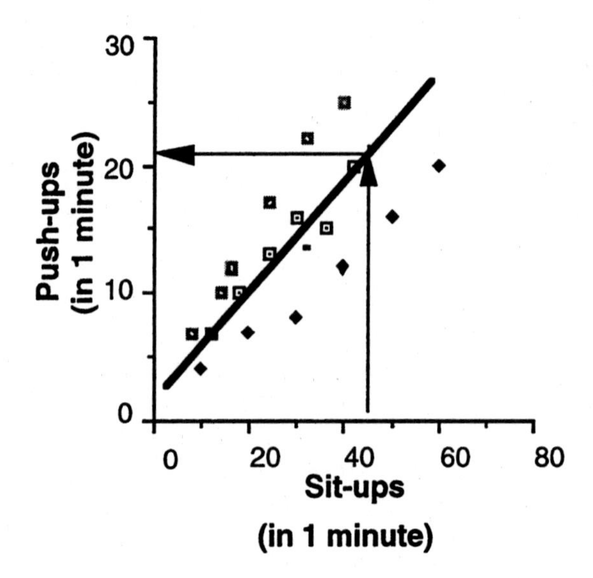

The **standard deviation** is a measure of dispersion. It quantitates how data points are distributed about the mean. Everyone has had experience with exams for which all students received approximately the same score. In this situation, the dispersion around the mean is very small; thus the standard deviation is small. On another exam, the scores may have ranged from excellent to dreadful. In this example, the dispersion around the mean is very large; thus the standard deviation is large. The standard deviation is used to estimate how much the individual measurements in a set of data deviate from the mean of the set. It takes into account all the measurements in the activity. The standard deviation may also be used to compare the degree to which two sets of data are equal.

Figure 4. Illustration of large and small standard deviation (SD) about the mean score on two exams

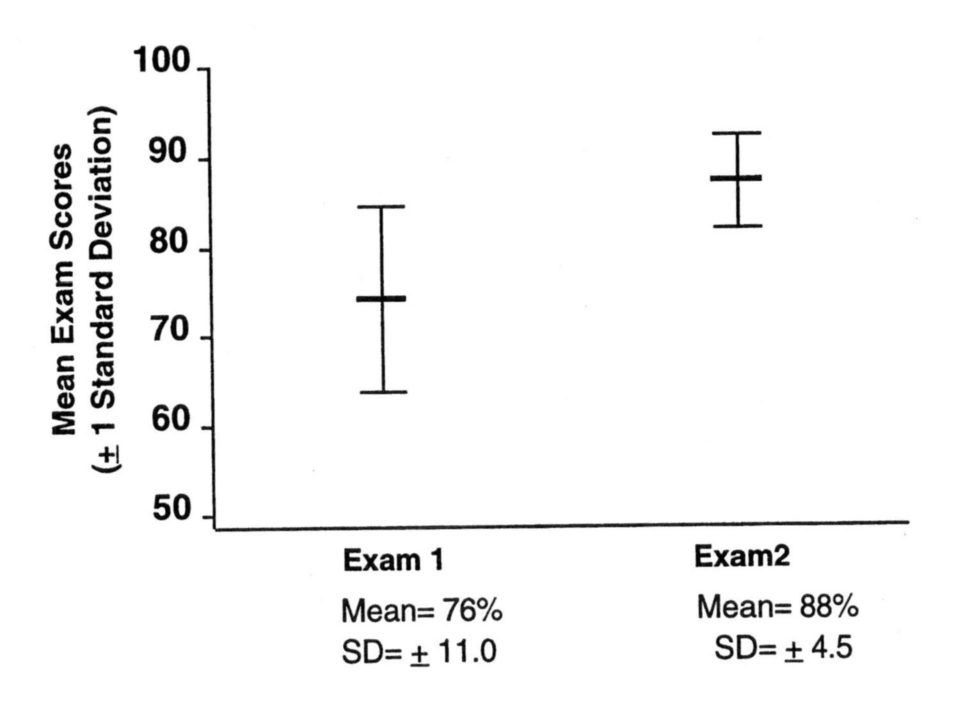

Part I: MEASURING LEG CIRCUMFERENCE

This exercise is designed to determine the mean, median, and mode of the circumference of each student's lower leg.

PROCEDURE

1. Divide the class into two groups by sex.

2. Further divide the two groups into sub-sets with a minimum of 3 members per team.

3. Obtain a string 1 meter in length

4. Have a student sit on the edge of a table so that his or her legs hang freely.

5. Hold the string horizontally behind the student's knee. Move the string up and down along the calf to determine the area of maximum circumference. Wrap the string around the calf and mark the string where one end of the string meets it. The distance between that end of the string and the point marked is the circumference of the leg. The string, or measuring tape, should be in contact with the skin, but it should not indent it. Record the measurement in Table 1.

6. Repeat step 5 two more times; having a different student obtain the measurement each time.

7. Using the caliper (see Appendix Figure 1), measure the skin-fold thickness on the medial side (internal) of the student's lower leg. Record the measurement in Table 1.

8. Repeat step 7 two more times, having a different student obtain the measurements each time. Were all three measurements the same? Explain.

9. Add the three values obtained for the leg and caliper measurements.

10. Divide each total by three to find the average or mean.

11. To find the **adjusted mean**, subtract the average caliper measurement from the average leg circumference measurement. The adjusted mean corrects for individual differences in body composition (muscle vs. fat).

12. Repeat steps 4 through 11 for each student in the class.

13. Record in Table 2 the adjusted mean values for all the males and all the females. Use these values to construct histograms in Graph 1 and Graph 2 for the males and for the females, respectively.

14. The highest bar on the histogram represents the most common measurement or the mode. Find the mode for each graph and record it on the line provided on page 11.

15. Find the median for each graph and record it in the space provided on page 11.

Table 1. Leg circumference and caliper measurements

	LEG CIRCUMFERENCE	CALIPER MEASUREMENT
MEASUREMENT #1	cm	cm
MEASUREMENT #2	cm	cm
MEASUREMENT #3	cm	cm
TOTAL =	cm	cm
MEAN =	cm	cm

Table 2. Adjusted mean leg circumference for each student

MALES	ADJUSTED MEAN	FEMALES	ADJUSTED MEAN

Males: mode = __________ **Males: median = __________**

Females: mode = __________ **Females: median = __________**

Graph 1. Histogram of Male Student Leg Circumferences

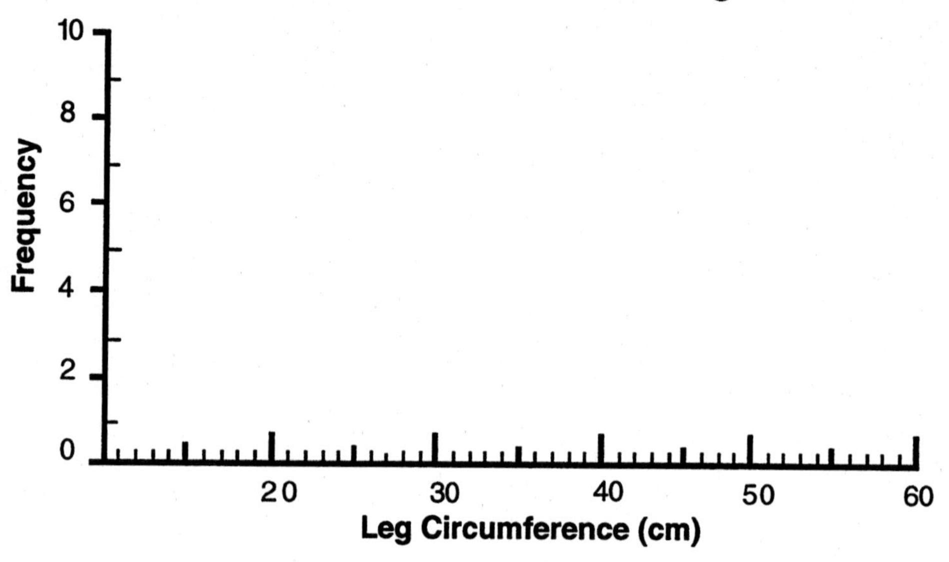

Graph 2 . Histogram of Female Student Leg Circumferences

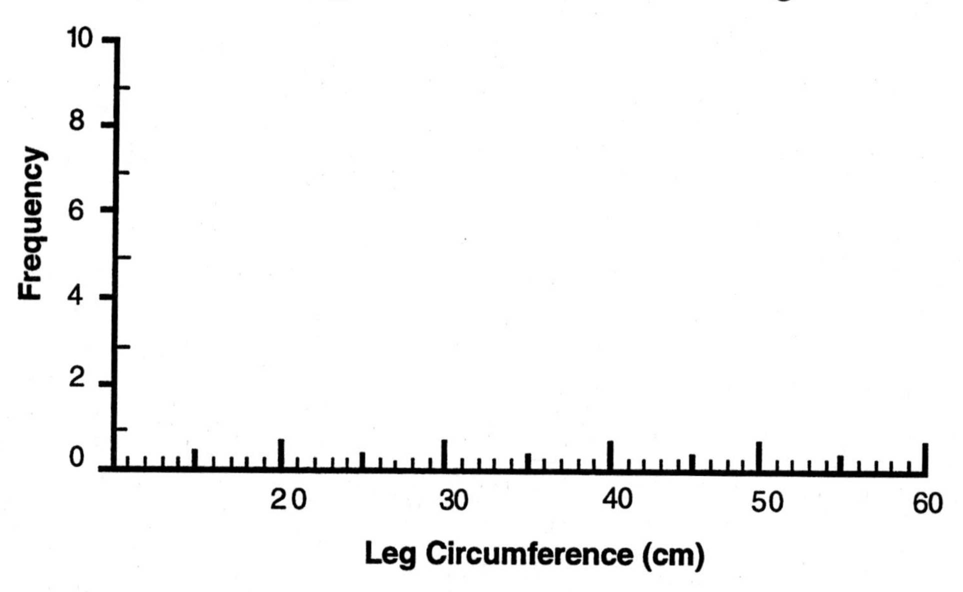

Part II: THE STANDING CALF JUMP

This exercise is designed to determine the relationship between a student's calf size and his or her vertical jump.

1. Mark a 1-meter strip of paper in 1-centimeter increments.

2. Tape the paper so that one end of it is at a point 2 meters above the floor.

3. Stand with the favored arm closest to the wall.

4. With feet flat on the floor, reach as high as possible and touch the paper strip; the highest point reached with the extended hand is the standing height. Record the height reached on the paper strip here: ______________ cm.

5. While keeping the trunk straight, bend at the knees and spring upward (jump) touching as high on the chart as possible; this is the jumping height. Record the height reached on the paper strip here: ___________ cm.

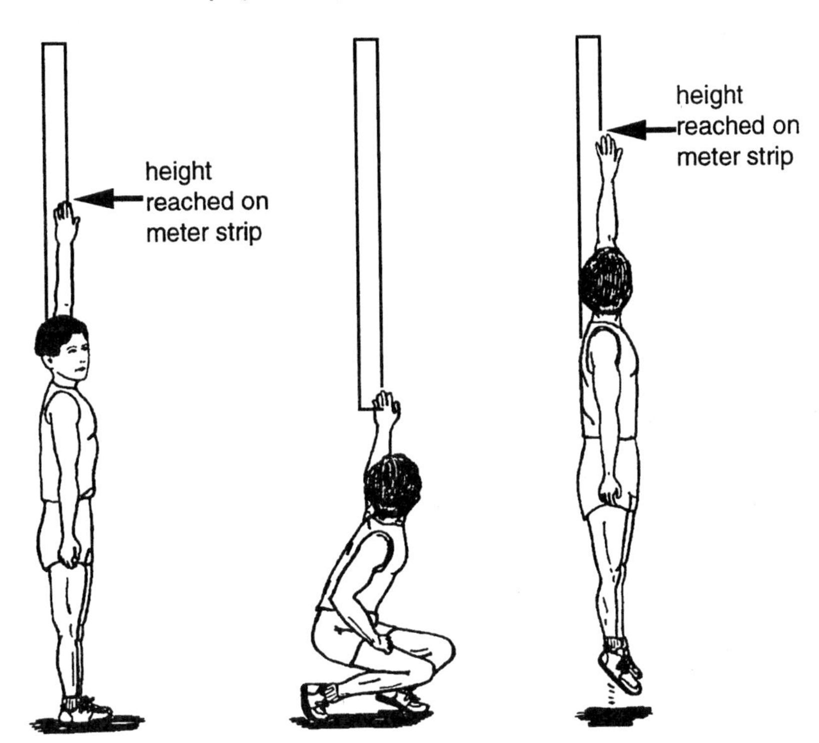

6. The difference between the two values represents the vertical height jumped. In Table 3 record the vertical height jumped by each student in the class.

7. On Graphs 3 and 4 plot the values obtained for the calf circumference and the vertical height jumped for each student. Calf size is on the abscissa, or *x*-axis, and vertical height jumped is on the ordinate, or *y*-axis.

8. Using a ruler as a guide, estimate the line of best fit or regression line through the scatter plots of Graphs 3 and 4. Draw the line with approximately the same number of points above and below the line. You will use the regression line to make predictions in Part III of this exercise.

Table 3. Vertical height jumped by each student in the class. (Remember to subtract the standing height from the jumping height to determine the standing vertical jump.)

MALES	HEIGHT JUMPED	FEMALES	HEIGHT JUMPED

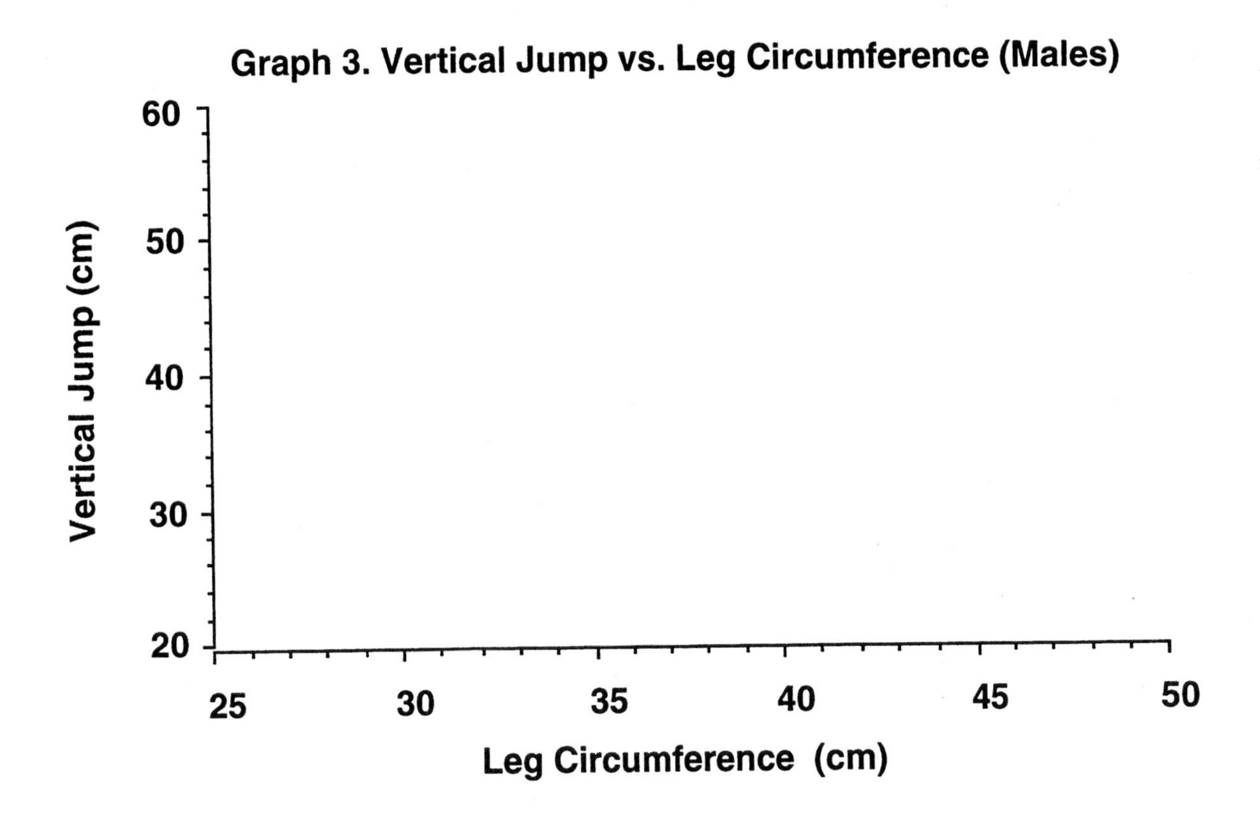
Graph 3. Vertical Jump vs. Leg Circumference (Males)
60
50
40
30
20
Vertical Jump (cm)
25
30
35
40
45
50
Leg Circumference (cm)

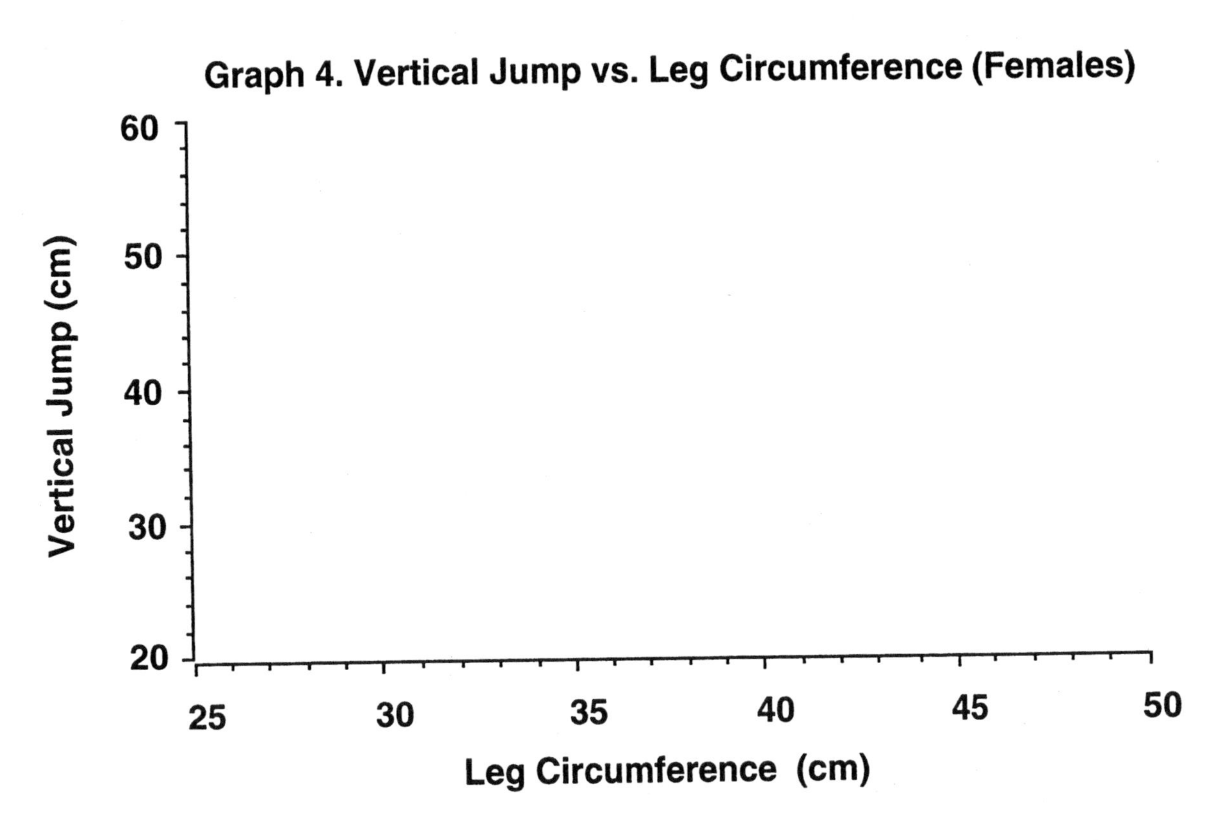
Graph 4. Vertical Jump vs. Leg Circumference (Females)
60
50
40
30
20
Vertical Jump (cm)
25
30
35
40
45
50
Leg Circumference (cm)

Part III: PREDICTING VALUES USING YOUR DATA

1. Measure the circumference of a student's lower leg and predict the vertical height that student can jump using your regression line. In order to do this, find the calf measurement on the *x*-axis. From there, move vertically until you cross the line of best fit (regression line). From that point, move horizontally to the left until you cross the *y*-axis. Read the value on the *y*-axis. This is the subject's predicted vertical jump height.

2. Now have the student jump. Compare the actual height jumped with the value predicted. (Note that you could also predict calf size from a measured jump.)

Part IV: STANDARD DEVIATION

1. Determine the mean of the jumping heights for both the males and the females.

2. In order to analyze the data, several statistical operations are necessary. The first is to determine the standard deviation for each of the means. The standard deviation is a measure of the dispersion of individual points around the mean. The standard deviation is defined as follows:

$$SD = \sqrt{\frac{\sum(X_i - \overline{X})^2}{n-1}}$$

Where SD = standard deviation
$\overline{X}$ = mean jumping height
X_i = each individual's jumping height
n = number of subjects
$\sum$ represents summation

3. In a normally distributed population, 68% of all individuals will fall within ± 1 standard deviation of the mean. For this experiment we will define the range as the mean ± 1 SD. For example, if the mean is 30 cm and the standard deviation is 4.5, the upper limit of the range is 30 + 4.5, and the lower limit is 30 - 4.5. Therefore the range of jumping heights is 25.5 cm to 34.5 cm. Determine the range for the males and for the females and record them here.

Males: range =__________ **Females: range =__________**

Figure 5. Sample range of vertical jumps for males and females

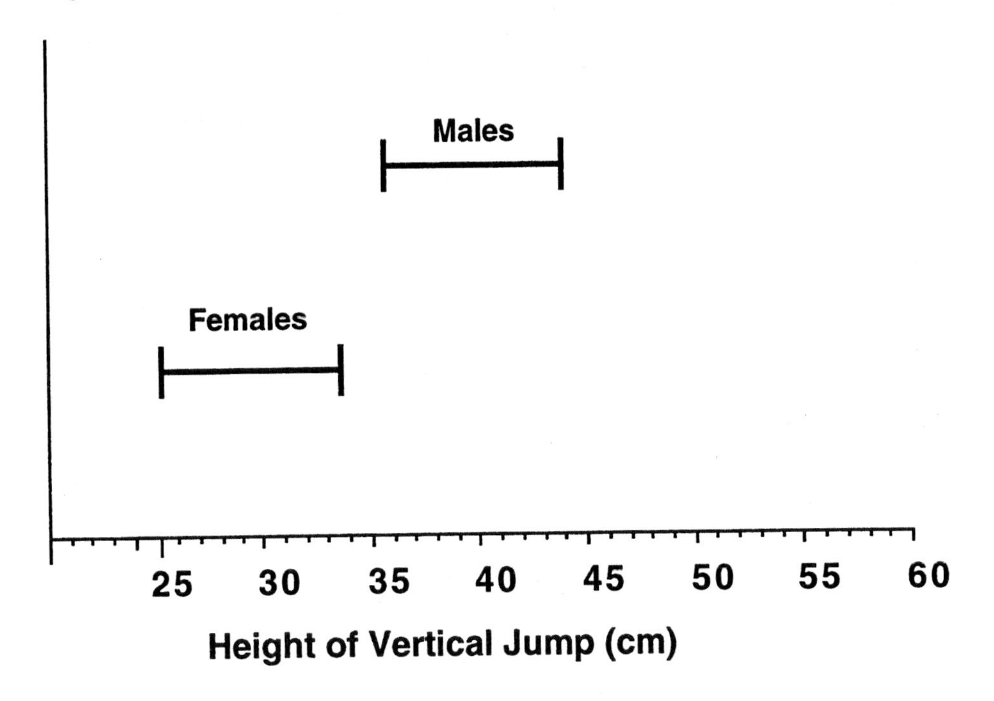

4. On Graph 5, produce a line for each gender whose length represents the range. If there is an overlap between the ranges of the two genders, then the difference is due to chance. If there is no overlap, then the difference is a significant finding. Figure 5 illustrates a sample range for females (25–35 cm) and a sample range for males (36–46). There is no overlap between the two ranges; therefore the difference between the mean for the males and the mean for the females is a significant finding. Another way to describe the results is to say that there is some reason for the difference in vertical jumps between males and females; perhaps biological or genetic factors are responsible.

Graph 5. Range for Male and Female Vertical Jumps

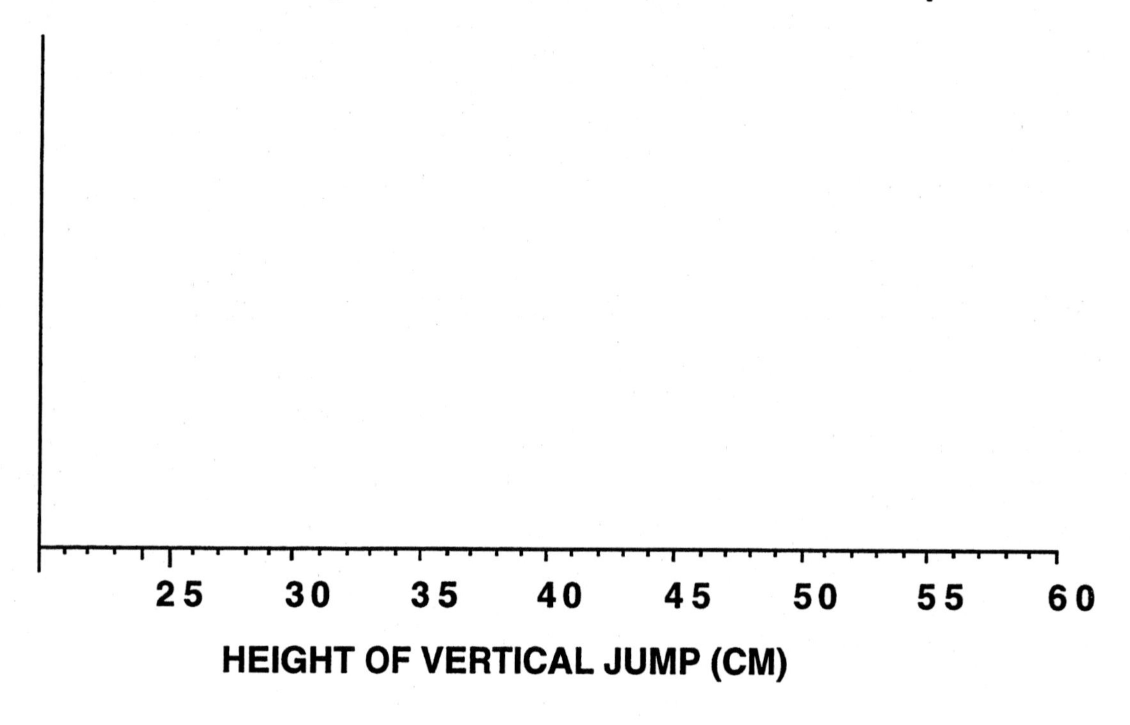

1. If you found no overlap, suggest a possible reason for the difference.
2. Did 68% of your student's scores fall within the range (mean $\pm$ 1SD)?

Part V: BODY RATIO VERSUS JUMPING HEIGHT

It is possible that the scatter plot data are random, so that the regression line is not easily drawn. A positive correlation between calf size and jumping ability would be expected, but this may not be true. To address this possibility, perform the following activity.

1. Record the body weight of each subject in kilograms in Table 4.

2. Formulate ratios by dividing the body weight by the calf circumference.

3. Plot the body weight–calf ratio versus jumping height for each student on Graph 6. (Notice that some people with different calf sizes are found to share the same ratios.)

Table 4. Body weight, leg circumference, and ratio of body weight to calf size for each student in the class.

MALES	BODY WT.	CALF SIZE	RATIO	FEMALES	BODY WT.	CALF SIZE	RATIO

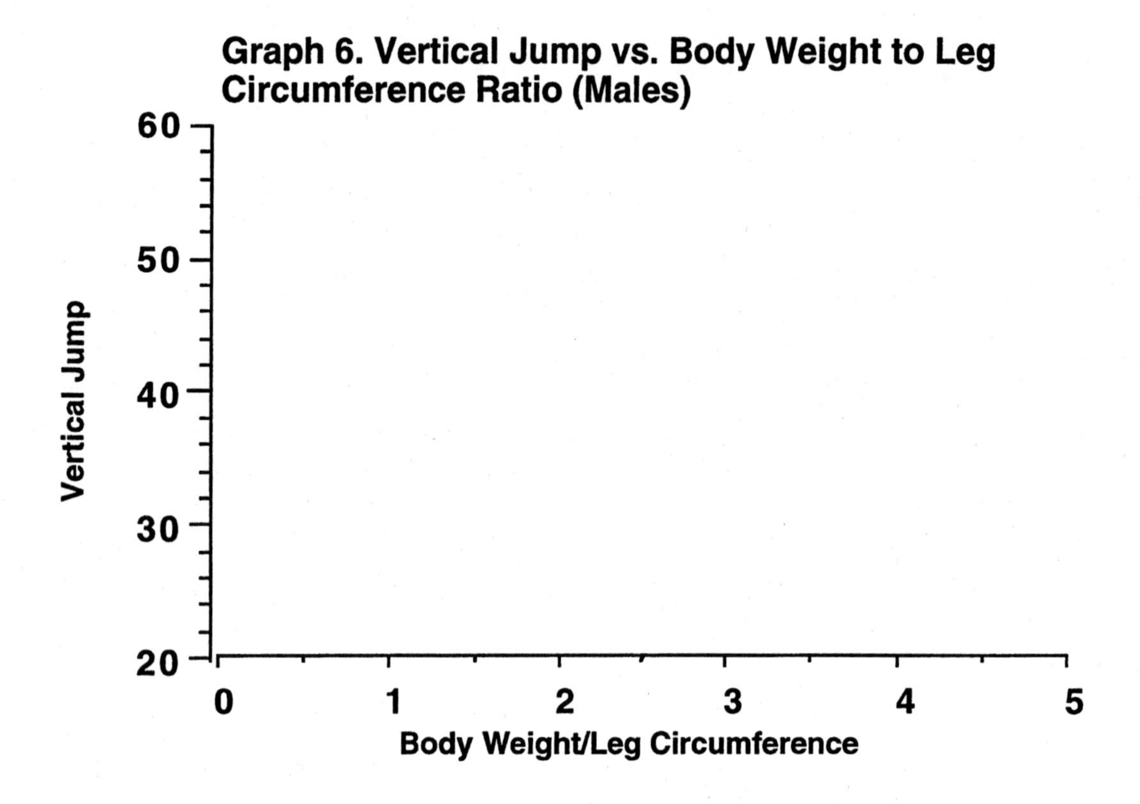

Graph 6. Vertical Jump vs. Body Weight to Leg Circumference Ratio (Males)

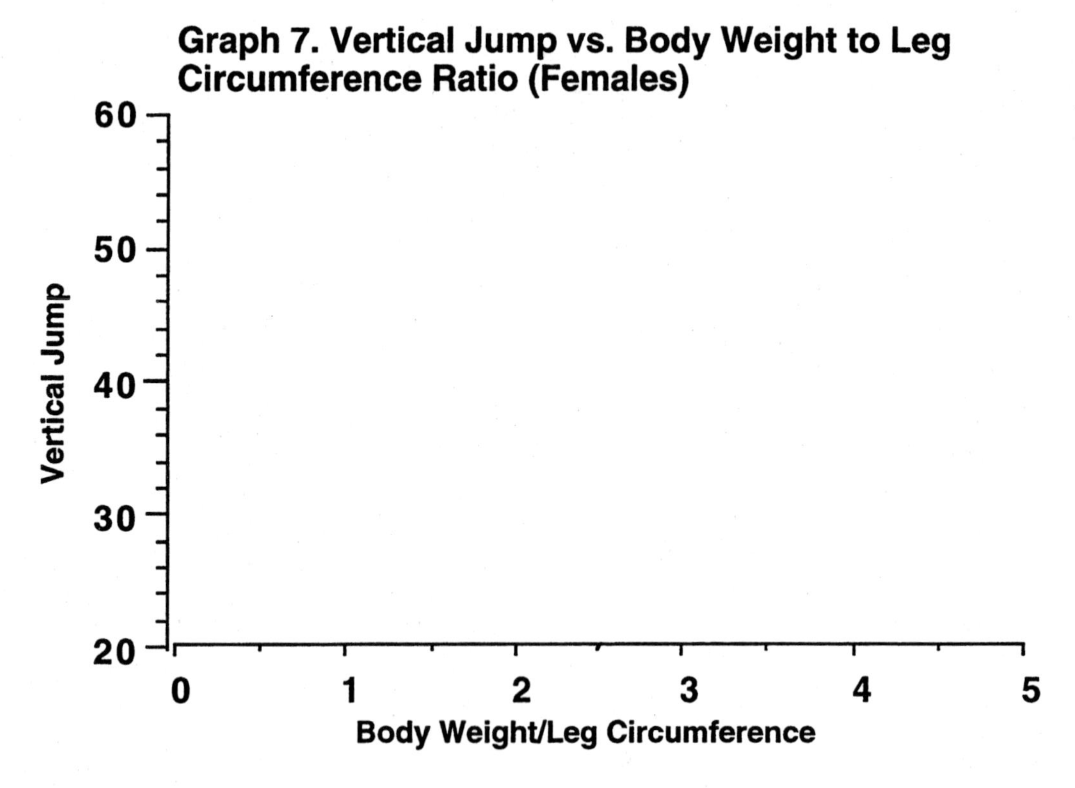

Graph 7. Vertical Jump vs. Body Weight to Leg Circumference Ratio (Females)

DISCUSSION

The differences observed in height jumped can be explained by examining some differences in muscle structure.

Each muscle is composed of two main fiber types that are responsible for muscle activity, fast-twitch (white) and slow-twitch (red) fibers. Fast-twitch muscle fibers, as their name suggests, are responsible for powerful, short activities, such as sprints. Slow-twitch fibers, in contrast, are essential for performing longer activities, for example, running cross country. A simple comparison is that between a duck and a rabbit. If you have ever eaten duck, you probably noticed that it is very oily. The oil is present because ducks use fats as a fuel source. The duck migrates long distances and needs more slow-twitch fibers for endurance. Slow-twitch fibers predominantly use *aerobic beta-oxidation* for energy. Aerobic refers to the presence of oxygen; beta-oxidation is the breakdown of fatty acids for energy. This process occurs in the mitochondria and requires the presence of oxygen. Muscles store oxygen in the form of myoglobin (a hemoglobin-like protein); therefore, there are more mitochondria in slow-twitch muscle fibers than in fast-twitch muscle fibers.

The rabbit, on the other hand, has mostly fast-twitch muscle fibers. Rabbits hop very quickly for a short distance and then stop to rest. Fast-twitch muscle fibers derive their energy from *anaerobic glycolysis.* Anaerobic refers to the absence of oxygen; glycolysis is the breakdown of glucose; therefore the fast-twitch muscle fibers contain high concentrations of glycogen, the stored form of glucose.

It is interesting to note that the difference between running aerobically and anaerobically is as simple as the difference between running a 400-m dash (anaerobic) and running an 800-m run (aerobic). During aerobic exercise, the oxygen in the inspired air provides the energy to complete the activity. Anaerobic activities are fueled *without utilizing the oxygen in inspired air.* A good sprinter can run the 400-m dash in under a minute. Fast-twitch muscle fibers provide rapid power for a period of a few seconds to a minute. Slow-twitch are able to contract from minutes to 2–3 hours.

The proportion of slow-twitch to fast-twitch fibers possessed by an organism is determined genetically, and the number of fibers is constant throughout the lifespan. The proportion of each type of fiber will, to a certain extent, determine a person's maximum physical capability in sports. For example, regardless of how hard a world-class sprinter trains for distance running, he or she will never become a top marathon runner. Some students who have the same body weight to circumference ratio as another student may jump higher simply because they were born with more fast-twitch fibers. See Table 5.

Table 5. Activity in Relation to Muscle Fiber Composition

ACTIVITY	FAST-TWITCH (%) (anaerobic)	SLOW-TWITCH (%) (aerobic)
Marathon	18	82
Swimming	26	74
Weight Lifting	55	45
Sprinting	63	37
Jumping	63	37

Note: The average male has a muscle fiber composition of 55% fast-twitch and 45% slow-twitch fibers.

QUESTIONS

1. **What exercises use mostly fast-twitch fibers?**
2. **What exercises use mostly slow-twitch fibers?**

Consult Table 6 to see if you are right.

Table 6. Activities and the Energy Systems They Employ

FAST-TWITCH (anaerobic)	SLOW-TWITCH (aerobic)
100-m dash	800-m dash
Jumping	200-m swim
Weight lifting	1500-m skating
Diving	Boxing
Football dashes	2000 m rowing
200-m dash	1500-m run
Basketball	400-m swim
Baseball home run	10,000-m skating
Ice hockey dashes	Jogging
Tennis	Marathon run
Soccer	Cross-country skiing

APPENDIX

Figure 1 Caliper

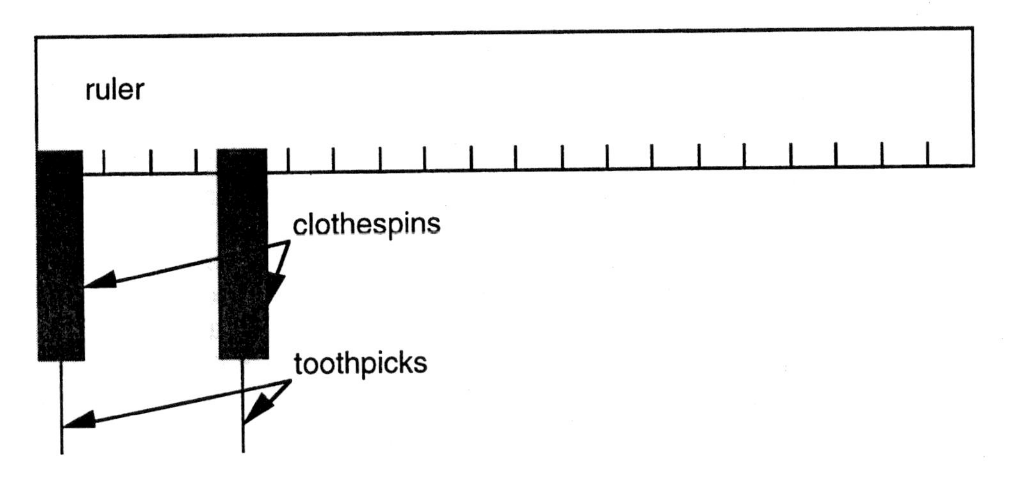

PROCEDURE

1. Place a toothpick on top of the gripping edge of a clothespin so that it extends about 3 cm beyond the edge of the clothespin. Fasten the toothpick to the clothespin with tape.

2. Repeat the process with the other toothpick and clothespin.

3. Clip the clothespins to the ruler. You may wish to anchor one clothespin to the ruler so that only one clothespin is moveable.

4. The measuring width can now be adjusted by sliding the clothespin along the ruler. This device will be used to measure the skin-fold thickness. Once you obtain a skin fold, position the calipers so the toothpicks extending from the clothespins are on either side of the skin fold. This may require two people. The distance between the two clothespins on the ruler will be the skin-fold measurement.

LABORATORY EXERCISE 2

THE SINGLE-CASE STUDY

BACKGROUND AND KEY TERMS

Classical experimental design requires a large sample size and both an experimental and a control group. The classical design has controls for time, sampling size, and measurement invalidity and is the ideal way to obtain significant experimental results. However, in specialized populations it may not be possible to measure a large number of subjects with similar characteristics. For example, imagine trying to study whether a treatment is effective for a high jumper who feels pain during an event. It would be difficult to find many people with the same problem. One way to get around this problem is by using the **single-case study**, which is also referred to as a **case study** or **clinical trial**. In a case study, the subject must serve as his or her own control.

The design used in the single-case study is the **A-B-A design**. The first component, A represents the baseline period, in which the individual is in a stable, unaltered condition. During the B period the treatment or intervention is introduced. The treatment is removed in the third period so that the individual may return to the baseline conditions (A). The return to baseline helps determine whether the response obtained during the B period (if there is one) is due to a preexisting condition or whether it is due to the treatment. The purpose of the single-case study is to establish a relationship between treatment and response. Figure 1 illustrates the pattern obtained in a typical study using the A-B-A design.

Figure 1. Expected results of an A-B-A design

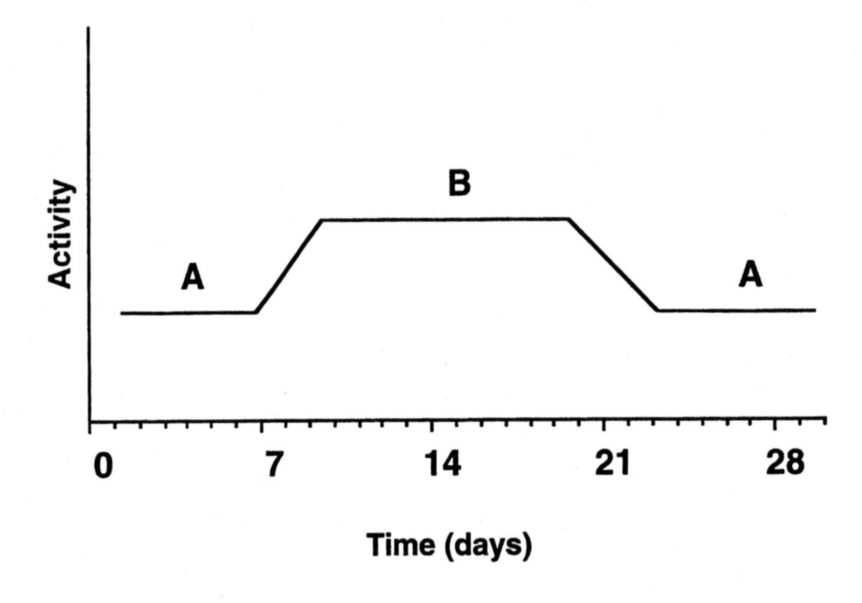

This exercise is designed to demonstrate a single-case study of a chosen activity.

CONDUCTING A SINGLE-CASE STUDY

PROCEDURE

1. Choose a preferred activity, and each day measure the designated parameter as specified within the activity. On days 1–7 measure the desired activity. On days 8–21, add the prescribed treatment or intervention. Remove the treatment on day 22; continue measurements until day 28.

2. Collect data from the measurements and record them in Table 1 (p. 31). Plot the data on Graph 1 (p. 32). The *x*-axis (independent variable) will always be time (days). The *y*-axis (dependent variable) will depend on the activity chosen. Lable the *y*-axis according to the instructions given within the activity.

3. In order to describe the effect the treatment has on the activity, calculate the **percent change** for the time between the initial baseline (A) and treatment periods (B) as well as between the treatment (B) and the final baseline periods (A). Calculate percent change as follows:

   a. Compute a mean for each of the three periods: the initial baseline (A), treatment (B), and final baseline periods (A).

   b. Substitute the computed means in the following equations:

$$\%\ \text{change} = \frac{\overline{T} - \overline{B}_i}{\overline{T}} \text{ X } 100 \quad \text{and} \quad \%\ \text{change} = \frac{\overline{T} - \overline{B}_f}{\overline{T}} \text{ X } 100$$

where $\overline{T}$ = treatment mean
$\overline{B}_i$ = initial baseline mean
$\overline{B}_f$ = final baseline mean

SUGGESTED ACTIVITIES

1. Biceps curl (Option 1)

On Graph 1, label the *y*-axis in centimeters in increments of 0.5 cm, and time (days) on the *x*-axis. With your arm hanging at your side, measure the circumference of your arm at the widest part of the biceps muscle. Record and plot daily measurements on the graph. Repeat this procedure each day for 28 days, always measuring the same arm. (Strength may also be determined. See p. 29 for determinants of strength.) This activity is designed to determine if strength training increases the circumference of the muscle.

(A) Days 1–7
Measure the circumference of your arm. Record and plot the results on the graph. These measurements establish the baseline period, A.

(B) Days 8–21
Begin a strength-training program of lifting a soup can or weight for three sets of 15 repetitions with the same arm. Begin the biceps curl with arm straight, hand at your side then bend the elbow and slowly lift the weight toward the shoulder. Measure the circumference of the arm after completing the set of exercises. Record and plot daily measurements. These measurements establish the intervention period, B.

(A) Days 22–28
Discontinue the strength-training program (biceps curls), but continue to measure the circumference of the arm. Record and plot the data on the graph. These measurements establish the second baseline period, A.

2. Biceps Curl (option 2)

This activity is identical with the previous activity, except that all measurements are made on the *nonexercising biceps* to determine if strength training with one limb improves the strength in the opposite limb. This experiment tests the cross-transfer theory. (This theory is explained on p. 30.) If you perform this activity, follow the same procedure as in option 1, but measure the circumference of the opposite arm.

3. Free-throws (requires basketball and hoop)

Stand at the foul line and attempt 10 free throws by tossing the basketball either underhand or overhead toward the hoop. Record and plot the number of free throws made each day on the graph. Label the y-axis in number of free throws made and the x-axis in days. This activity is designed to determine if the number of free throws made improves with more practice.

(A) Days 1–7
Record the number of free-throws made out of 10 attempts once each day. These measurements establish the baseline period, A.

(B) Days 8–21
Record and plot the number of completed free-throws out of 10 attempts each day. Take an additional 50 practice shots each day as well. These measurements establish the intervention period, B.

(A) Days 22–28
Discontinue the 50 practice shots each day. Continue to record and plot the number of free-throws made out of 10 attempts. These measurements establish the second baseline period, A.

4. Flexibility

While sitting on the floor with your legs extended and toes pointing straight up, try to touch your toes with outstretched fingers. Have someone measure the distance between your fingers and toes. Record the distance reached in front of your toes as a negative number, and the distance reached beyond your toes a positive. These measurements determine flexibility. Measure flexibility daily; record and plot the data on the graph. Let the *y*-axis be flexibility measured in centimeters, and the *x*-axis, days. This activity is designed to determine if flexibility increases with stretching.

(A) Days 1–7
Measure flexibility each day. Record and plot the data on Graph 1. These measurements establish the baseline period, A.

(B) Days 8–21
Measure flexibility each day. Following the measurement, continue stretching for 10 minutes. The stretch should consist of slowly bending from the waist down toward your knees; do not bounce or bend your knees. These measurements establish the intervention period, B.

(A) Days 22–28
Discontinue the straight-leg stretches but continue measuring flexibility. These measurements establish the second baseline period, A.

5. Flexed-Arm Hang (requires a chin-up bar)

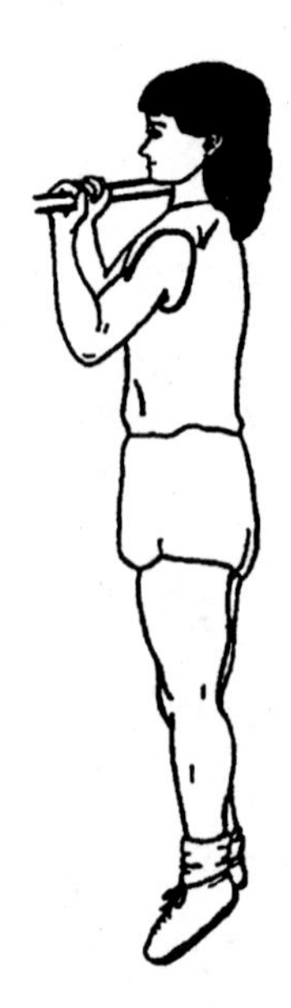

Measure the time (seconds) that you are able to hang from a chin-up bar; your arms should be flexed and your chin must remain above the bar. This activity is designed to determine if performing numerous flexed-arm-hang tests increases the time you are able to hang with your chin above the bar. Label the *y*-axis in seconds and the *x*-axis in days.

(A) Days 1–7
Record and plot the number of seconds you are able to maintain your chin above the bar. These measurements establish the baseline period, A.

(B) Days 8–21
Perform the flexed-arm hang daily; record and plot the number of seconds completed on the graph. Perform five additional flexed-arm hangs each day. Record the time from only the first flexed-arm hang each day. These measurements determine the intervention period, B.

(A) Days 22–28
Perform only one flexed-arm hang each day. Record and plot the time on the graph. These measurements establish the second baseline period, A.

6. Push-Ups

Record the maximum number of push-ups you are able to complete in 1 minute. Perform push-ups the following way:

Lie face down on the mat and place your hands palm down on the mat next to your shoulders. Push yourself up so that your arms are fully extended. Return to the down position, allowing your chest to touch the mat. Be sure to keep the back, hips, and legs in the same plane throughout the exercise.

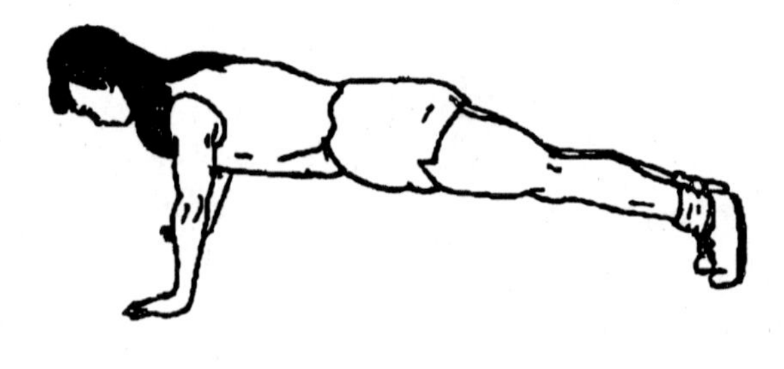

This activity is designed to determine if performing three sets of push-ups a day increases the maximum number of push-ups you can complete in one minute. Let the *y*-axis be number of push-ups completed and the *x*-axis, days.

(A) Days 1–7
Record the maximum number of push-ups completed in 1 minute. These measurements establish the baseline period, A.

(B) Days 8–21
Record the maximum number of push-ups completed in 1 minute. Perform an additional three sets of 10 push-ups at the end of the day. These measurements establish the intervention period, B.

(A) Days 22–28
Discontinue performing the three sets of 10 push-ups each day. Continue to record the number of push-ups you can complete in 1 minute. These measurements establish the second baseline period, A.

7. Ten-Repetition Maximum (to determine strength)

Using weights, perform a biceps curl. Begin with the arm straight, hand at the side, then curl the arm up until it is fully flexed. Estimate the maximum weight that you can lift 10 consecutive times. First, test the right arm. Start at a weight below your maximum and lift the weight 10 times. Increase the weight in increments until the maximum weight is reached. For instance, add a pound each time. Allow 5 minutes of rest in between repetitions for the muscle to recover. The "max" is the maximum weight that can be lifted 10 times. Repeat the procedure for the left arm. Record the maximum weight lifted each day. Divide the "Measurement" column of Table 1 into two parts and record the max of the left arm in the left half and the right arm in the right half of the column. Label the *y*-axis of your graph in kilograms and the *x*-axis in days.

(A) Days 1–7
Record the 10-repetition maximum weight lifted each day for both arms. These measurements establish the baseline period, A. Plot the measurements for each arm on Graph 1. Use different symbols for the right and left arms.

(B) Days 8–21
At the beginning of each day record the maximum weight lifted by each arm. Later in the day complete three sets of 10 repetitions with only the right arm; the first set doing at 50% of maximum, the second set at 75% of maximum, and the final set with the maximum weight. For example, if your 10 repetition maximum is 20 kg, you would first lift 10 kg, then 15 kg and finally 20 kg. When looking for a maximum, remember to always load the muscle as much as possible. The maximum may vary from day to day. These measurements establish the intervention period, B.

(C) Days 22–28
Discontinue performing the additional three sets per day. Measure the 10-repetition maximum daily, and record the data in Table 1. These measurements establish the second baseline period, A.

CROSS-TRANSFER THEORY

Cross-transfer theory states that muscles that are not being exercised also increase in strength. This theory refers specifically to the analogous muscles on the side of the body opposite that whose muscles are being exercised. In other words, if the right arm is trained, the left arm will also increase in strength. Why?

Muscles increase in strength through two processes. The first is probably the most well known: **hypertrophy**, in which muscle fibers enlarge with repeated use. There is no increase in the number of fibers, so the size increase is due to enlargement of the muscle fibers. For a muscle to increase in strength, it must lift a load; that is why weight lifters have large, well-defined muscles.

However, an increase in strength occurs before there is a noticeable increase in muscle size due to changes in the brain or the nervous system. Usually the brain can recruit only a percentage of the muscle fibers for the action that it wants to perform. With an increased load on the muscle, the brain is able to recruit a larger percentage of the available fibers. Recruitment of more fibers increases the strength of the muscle.

As the cross-transfer theory states, there is a strength increase in the unexercised muscle. It seems that the brain is able to recruit more fibers in both limbs because the body is always trying to maintain a steady state or remain balanced. In this case, the brain sends nerve signals to both arms. However, some people do not have the same build on each side of their body. For instance, a pitcher has a larger pitching arm. Initially both arms gained in ability to recruit fibers; however, the exercised arm continued to increase in size (hypertrophy).

Table 1. Daily measurements taken for 28 days. Days 1–7 correspond with period A, days 8–21 correspond with the B period, and days 22–28 correspond with the second A period.

DAY	Measurement	DAY	Measurement
1		15	
2		16	
3		17	
4		18	
5		19	
6		20	
7		21	
8		22	
9		23	
10		24	
11		25	
12		26	
13		27	
14		28	

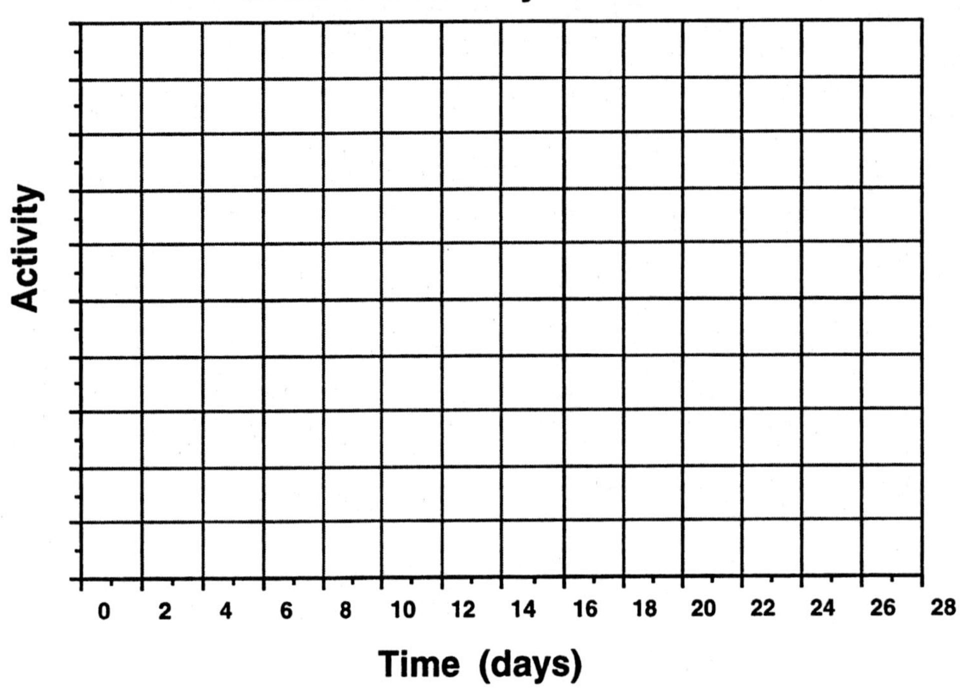

ANALYSIS OF DATA

1. What do you expect the graph to look like if there is a relationship between treatment and response? What did you find?

2. Is there a relationship between your activity and the treatment?

3. What was the percent change between the initial baseline and treatment periods?

4. What was the percent change between the treatment and final baseline periods?

5. What are some problems with performing, analyzing, or making assumptions from a single-case study?

LABORATORY EXERCISE 3

CARDIOVASCULAR RESPONSE TO EXERCISE

BACKGROUND AND KEY TERMS

During exercise, the tissues (skeletal muscle) require additional nutrients (e.g., oxygen, glucose, fats) to perform their functions. Since oxygen is required to perform work, oxygen consumption is often used as a measure of exercise intensity. Oxygen and nutrients are carried by the blood. Therefore, to meet tissue needs, the heart must pump more blood than it does at rest. There are two ways that the heart can deliver more blood or increase the **cardiac output (CO)**. The first is to increase the rate at which the heart beats, and the second is to increase the amount of blood pumped in a given beat.

Heart rate (HR) is defined as the number of times the heart beats in 1 minute. One can detect heart rate by palpating (touching) a peripheral artery and counting the beats. For example, using the index and middle fingers, feel your pulse on either the distal portion of your radius (thumb side of forearm close to wrist) or the carotid artery (natural groove between the neck muscle and the throat).

Immediately at the onset of exercise, the nervous system quickly causes the heart to beat faster. This is followed by a slower increase in HR as blood flow matches itself to tissue needs. This is depicted graphically in Figure 1.

Figure 1. Heart rate response to exercise

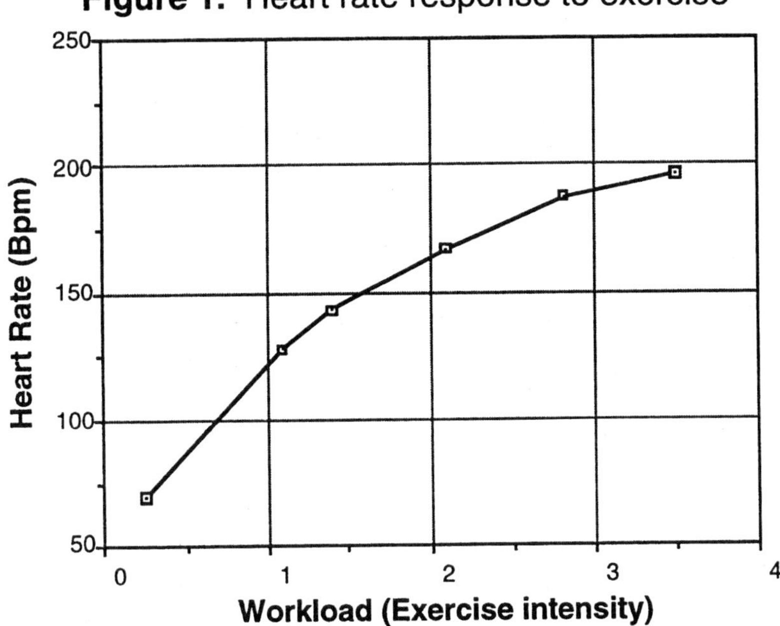

Stroke volume (SV) is defined as the amount of blood that the heart pumps in a given beat. Note that the heart contains approximately 150 milliliters (ml) of blood before it beats. The period of time before the heart beats, while the heart is filling with blood, is called *diastole.* When the heart contracts, blood is forced out of the heart and the volume drops to approximately 70 ml; this period is known as *systole.* The difference between these two volumes, the stroke volume, is approximately 80 ml of blood. As with heart rate, at the beginning of exercise the nervous system increases the SV, and there is a slower rise in SV later with exercise. The SV response to exercise is depicted in Figure 2.

Heart fills with blood during diastole.

Ventricles of the heart contract and blood is ejected during systole.

Figure 2. Stroke volume response to exercise

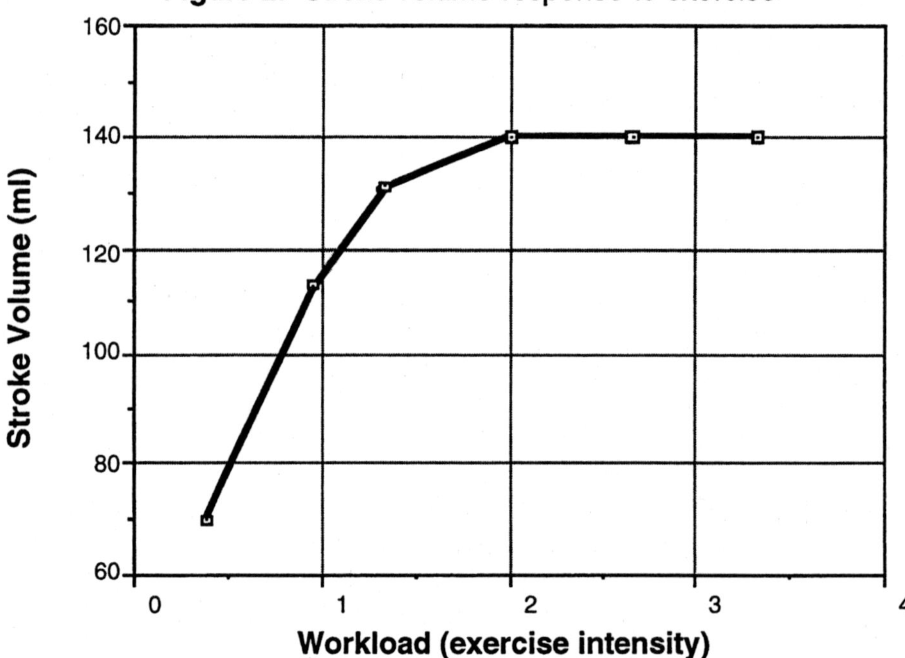

Cardiac output (CO) is defined as the amount of blood that the heart pumps to the tissues each *minute*. CO increases with exercise because both HR and SV increase. This relationship is expressed by the following formula:

CO = SV X HR

where CO = cardiac output (L/min)
SV = stroke volume (ml/min)
HR = heart rate (bpm)

Using Figures 1 and 2 to calculate CO, and plot the results in Figure 3.

Figure 3. Cardiac Output Response to Exercise

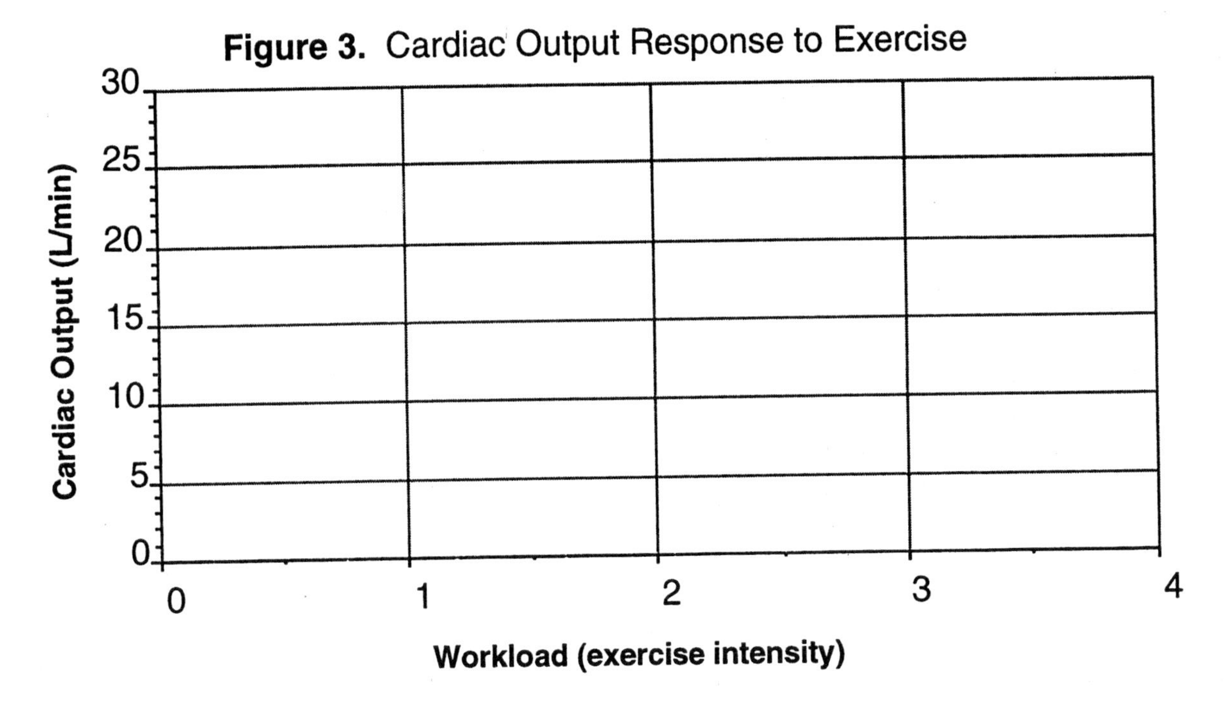

You may be surprised that CO increases to the same extent at the same exercise level, independent of fitness level. For example, a world-class marathon runner (Joan Benoit Samuelson) will have the same CO as Ms. Couch Potato at a set work level. This clearly is a surprise and raises the following question:

"What is the sense of exercising?"

Well, although one of the world's greatest marathoners and Ms. Couch Potato have the same CO, the marathoner achieves her CO with a much lower HR and a much larger SV.

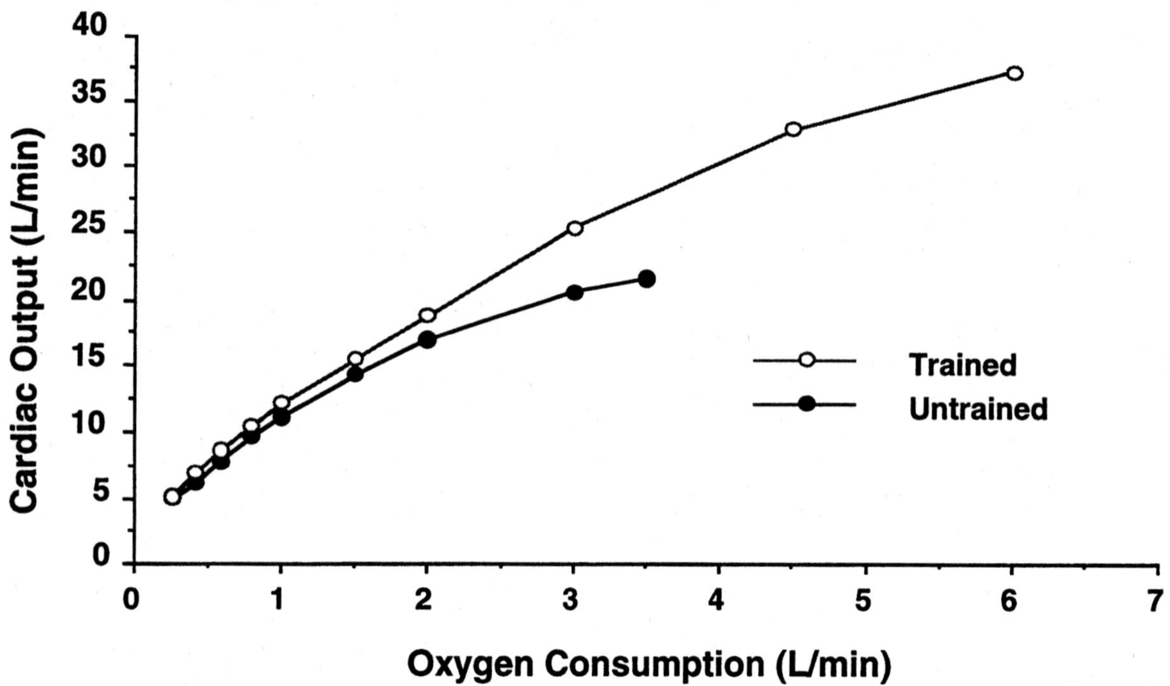
Figure 4. Cardiac Output Response to Exercise
Cardiac Output (L/min)
40
35
30
25
20
15
10
5
0
0
1
2
3
4
5
6
7
Oxygen Consumption (L/min)
Trained
Untrained

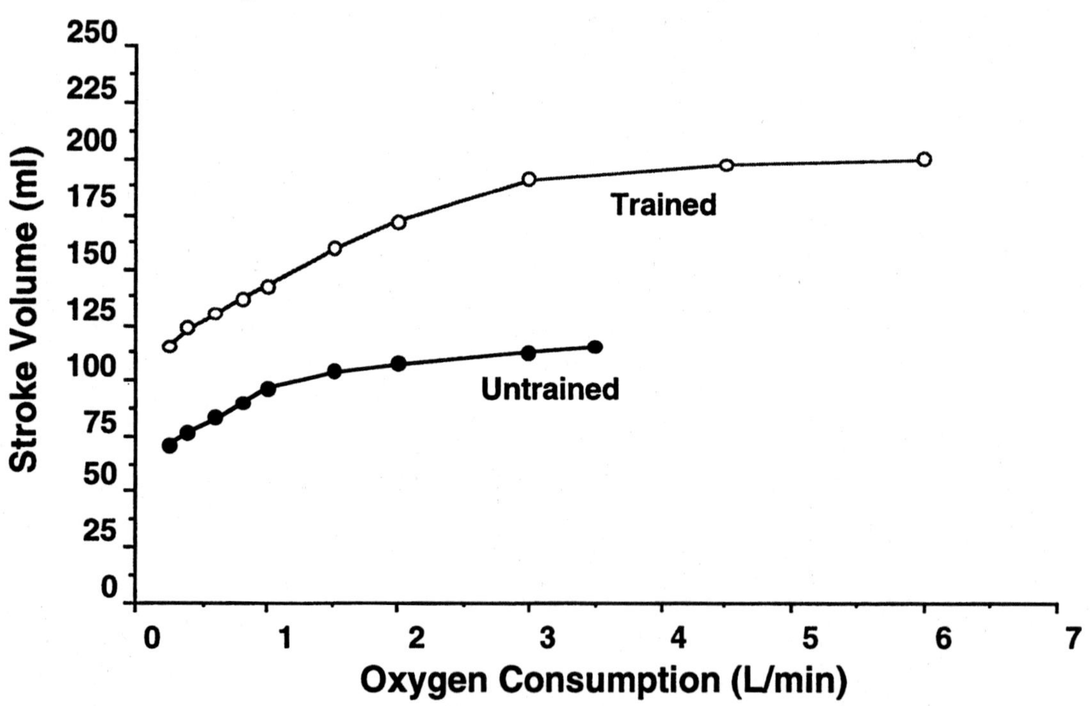
Figure 5. Stroke Volume Response To Exercise
Stroke Volume (ml)
250
225
200
175
150
125
100
75
50
25
0
0
1
2
3
4
5
6
7
Oxygen Consumption (L/min)
Trained
Untrained

Figure 6. Heart Rate Response To Exercise

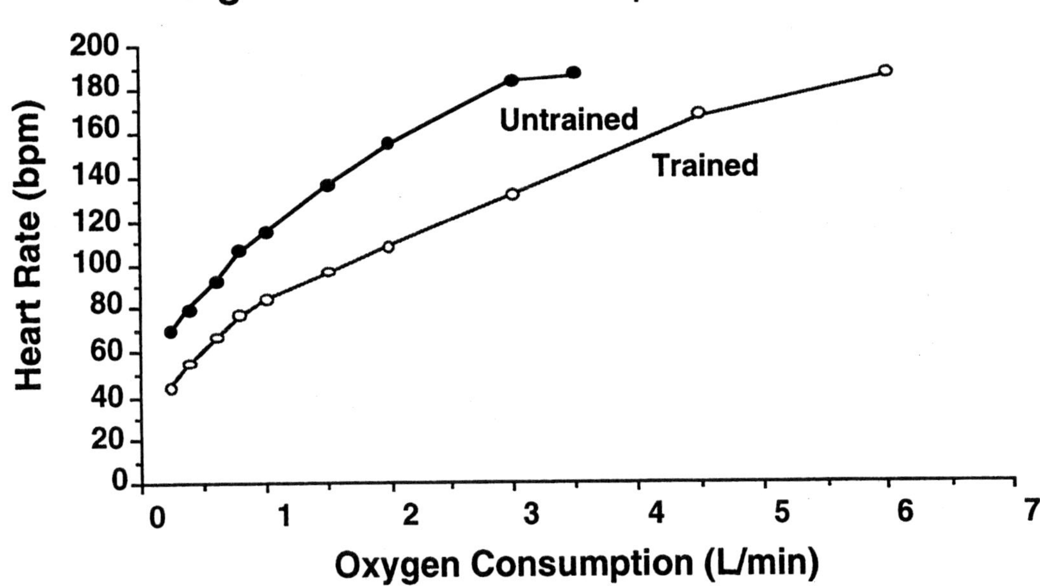

It is worth repeating that the means by which the trained runner and the untrained couch potato increase their CO is different. The CO response to exercise in a trained runner and untrained individual is depicted graphically in Figure 4. Recall that CO = SV X HR. The trained person will achieve the same CO with a lower HR because of a larger SV, while the untrained person has a higher HR and a smaller SV. These differences are depicted in Figures 5 and 6.

These figures also illustrate another advantage that exercise training offers. As mentioned, the marathoner achieves the same CO at a lower HR. Also notice that the maximum achievable heart rate is the same for both the trained and the untrained person. However, the trained person has a much lower resting heart rate. This means that the trained person has a greater HR reserve (difference between HR at rest and the maximum HR). This translates into a higher maximum CO for the trained individual. Consequently, this person is capable of performing at a higher level of exercise than that attainable by the untrained person.

The step test uses the preceding information as the basis for evaluating cardiovascular fitness. The HR response for a fit individual will be lower during and after exercise due to the adaptations created by training. By measuring the HR immediately following exercise, one can compare the results with information gathered from many trained and untrained people in the past (normal data) and thus allows one to predict the fitness level of people currently being tested.

Part I: THE 3-MINUTE STEP TEST

This activity is designed to determine the fitness level of the students in the class. To evaluate cardiovascular fitness, you will need to perform a short exercise activity (3 minutes of stepping). You will take heart rate measurements before the step test (at rest) and immediately following the exercise. A comparison of your heart rate values with standardized Tables 1a and 1b will determine your fitness rating.

PROCEDURE

The 3-minute step test requires a step bench, metronome, and timer. It is conducted in the following manner:

1. While sitting, obtain a 1-minute pulse. Use your fingers and not the thumb to feel the pulse on the radial (thumb side) portion of the wrist or on the neck at the carotid artery. Count the number of beats in 1 minute. Be careful when obtaining heart rate from the carotid artery (neck). If you apply too much pressure on the carotid artery, reflex slowing of the heart rate will occur. Record this as resting heart rate in Table 2.

2. Face the step bench and practice stepping up and down until you are comfortable with the sequence.

3. You must step 24 times each minute. To maintain a steady rate of stepping, a metronome is set at 96 beats per minute. Therefore, each step will consist of four beats in the following pattern:

   Beat 1: step up with leading foot
   Beat 2: step up with other foot
   Beat 3: step down with leading foot
   Beat 4: step down with other foot

4. The teacher or your partner should start the metronome and the timer once you are ready. The timer should be set for 3 minutes.

5. Step with the beat of the metronome for 3 minutes. It may be helpful to say "up, up, down, down" with each beat of the metronome.

6. With 15 seconds left, locate your pulse and prepare to take a measurement when indicated by the teacher.

7. The teacher or your partner should call out the last step as "last step, up, up, down, and sit."

8. At the end of 3 minutes, sit down and **immediately** (within 5 seconds) take a 1-minute pulse. Record this as the final heart rate in Table 2.

9. To determine your fitness rating, compare the resting and final heart rates with those in Tables 1a and 1b. Record your fitness rating in Table 2.

Table 1a. Fitness ratings for males based on HR

Rating	% Ranking	Resting HR	Final HR
Excellent	100	49	70
	95	52	72
	90	55	78
Good	85	57	82
	80	60	85
	75	61	88
Above Average	70	63	91
	65	64	94
	60	65	97
Average	55	67	101
	50	68	102
	45	69	104
Below Average	40	71	107
	35	72	110
	30	73	114
Poor	25	76	118
	20	79	121
	15	81	126
Very Poor	10	84	131
	5	89	137
	0	95	164

Table 1b. Fitness rating for females based on HR

Rating	% Ranking	Resting HR	Final HR
Excellent	100	54	72
	95	56	79
	90	60	83
Good	85	61	88
	80	64	93
	75	65	97
Above Avg.	70	66	100
	65	68	103
	60	69	106
Average	55	70	110
	50	72	112
	45	73	116
Below Avg.	40	74	118
	35	76	122
	30	78	124
Poor	25	80	128
	20	82	133
	15	84	137
Very Poor	10	86	142
	5	90	149
	0	100	155

Table 2. Heart rate measurements obtained before (resting) and after (final) the 3-minute step test

RESTING HEART RATE	FINAL HEART RATE	FITNESS RATING

ANALYSIS OF PART I

1. How do the athletes in the class compare with nonathletes with respect to their resting heart rate and the final heart rate?

2. What is the fitness rating of most athletes in class?

Part II: HEART RATE RESPONSE TO UPRIGHT POSTURE

"It certainly was a bold enterprise of nature to create quadrupeds like man or the giraffe with a predominantly vertical extension, who carry their heads and hearts at a considerable distance above the center of gravity of the body." (Gauer and Thron, 1965).

"If a careful engineering analysis of the aeronautical features of the bumblebee leads to the conclusion that these insects cannot fly, then a hydrodynamic analysis of the human circulation might lead to the conclusion that people cannot stand up." (Rowell, 1986)

When the body is in a supine position (lying on the back), a large volume of blood is concentrated in the thorax. When the body assumes an upright posture, gravity pulls the blood toward the legs, leaving much less blood in the thorax. The initial effect is a reduction of stroke volume and blood pressure. This is followed by a nervous system reflex that restores blood pressure by constricting peripheral blood vessels and increasing heart rate.

One physical effect of a lower blood pressure is the feeling of light-headedness. The intensity of this symptom is related to the degree of a person's physical fitness. One would expect a trained person to have fewer such problems than an untrained person. However, the opposite effect is actually observed. What can account for this? The answer is related to the amount of time it takes for the heart to adjust to the stress associated with a change in body position. The shorter the amount of time, the fewer the symptoms. The cardiovascular systems of those who are physically fit take longer to adjust than those who are not. The mechanism responsible for the delayed response is not known.

This exercise is designed to demonstrate the effects of changing one's posture on heart rate.

PROCEDURE

1. Work in teams of three. One person will be the subject, another will do the timing and recording, the third will take the pulses.

2. While the subject is lying supine on the floor, record his or her heart rate for 1 minute. Palpate the radial or carotid pulse with fingers and count the number of beats in 1 minute. Enter this value under the "rest" column in Table 3.

3. As the subject rises to a standing position, start the timer and take the heart rate for 1 minute for the next 5 minutes. Record the values in Table 3.

4. Change roles and repeat the experiment until each student has been the subject.

Table 3. Heart rate response to standing

NAME	REST	1 min	2 min	3 min	4 min	5 min

5. If, in Part I, any of the students had a fitness rating that was *above average* from Table 2 (Excellent, Good, Above Average) or *below average* (Very Poor, Poor, Below Average), then continue this activity by entering the data for these students in the appropriate Table (4a or 4b).

6. Compute the mean and standard deviation for each column, using the same formula in Exercise 1, p.16.

7. Using the calculated means, graph the above average and the below average data on Graph 1.

ANALYSIS OF PART II

1. How do the standing heart rates in the above average fitness groups differ from those in the below average groups?

2. Which group (above average or below average) reported feeling the most dizziness?

3. What are the physiological advantages and disadvantages of being in the above average category?

Table 4a. Above Average

NAME	REST	1 min	2 min	3 min	4 min	5 min
TOTAL =						
MEAN =						
STD DEV =						

Table 4b. Below Average

NAME	REST	1 min	2 min	3 min	4 min	5 min
TOTAL =						
MEAN =						
STD DEV =						

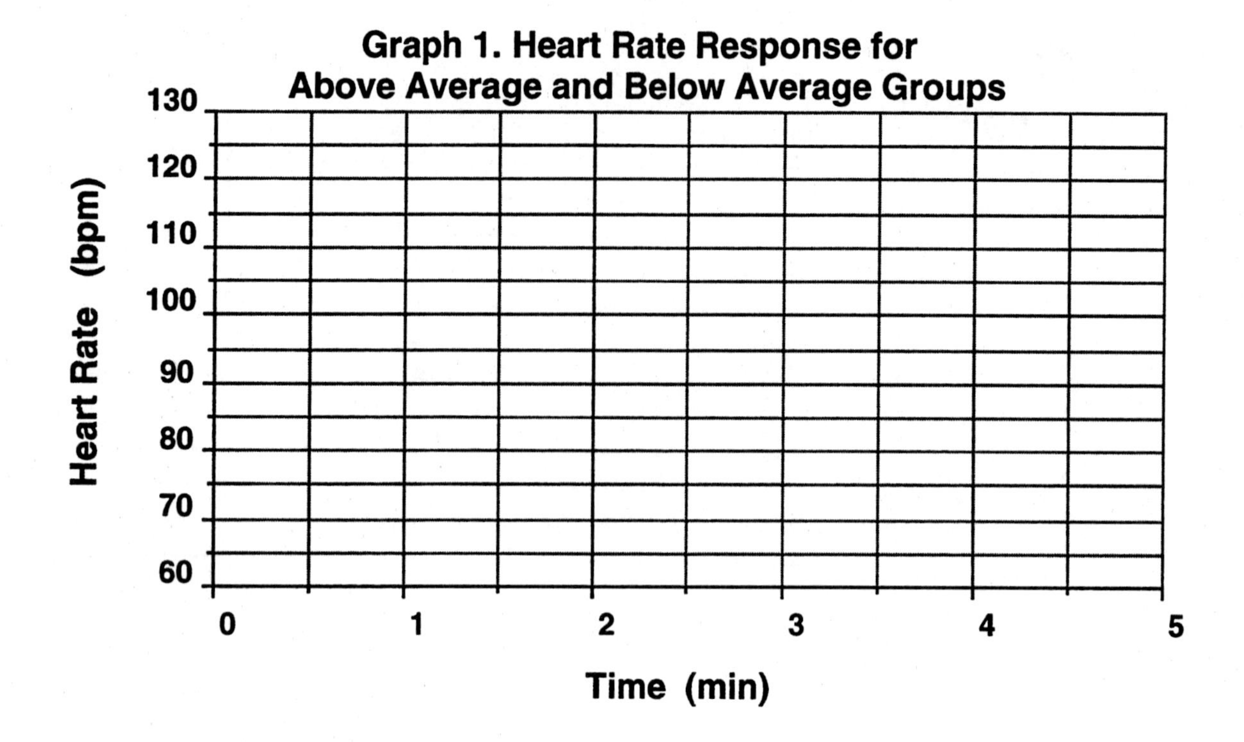

POINT OF INTEREST

The activities of this laboratory exercise relate to a similar phenomenon that occurs in astronauts as they reenter the earth's atmosphere from space. The weightlessness of space causes the blood to pool in the chest area in the same manner that lying down does. Upon reentry, gravity pulls the blood away from the head toward the lower extremities in the same way that standing up pulls the blood away from the head. As a result less blood goes to the brain, so the astronauts have a tendency to pass out. Because they are in excellent physical condition, the response is greatly enhanced. Likewise, those students who are in better physical condition should experience more problems than those who are not.

LABORATORY EXERCISE 4

RESPIRATORY RESPONSES

BACKGROUND AND KEY TERMS

As you sit and breathe normally, the air you are taking in and out is your **tidal volume (TV)**. This is the amount of air that normally exchanges oxygen in your lungs. If you take a deep breath, you notice that you are forcing yourself to take in more air than you did before when you were breathing normally. This "extra" air beyond the tidal volume is the **inspiratory reserve volume (IRV)**. Analogous to the IRV, there is an **expiratory reserve volume (ERV)**, which is the "extra" air breathed out beyond normal expiration. If you breathe out as much as you can, you are breathing out your tidal volume plus your expiratory reserve volume. The reserve volumes are very important, for they serve as a backup. Under normal conditions, these volumes are not needed; however the reserve volumes are used by the body to adapt to stressful conditions. It is obvious that more oxygen is required to feed tissues when one is playing soccer or tennis than when one is sleeping. The cardiovascular and respiratory systems work in concert to deal with these differences in physical activity. It is impossible to breathe out all the air in our lungs no matter how hard we try; the air remaining in the lungs is called the **residual volume (RV)**.

Capacity refers to how much a container can hold. If we imagine our lungs as containers, four capacities can be described. The **total lung capacity (TLC)** is the maximum amount of air that the lungs can hold. The **vital capacity (VC)** is similar to the TLC but excludes the residual volume. The vital capacity is the maximum amount of air that can be expired, following a maximal inspiration. The amount of air in the lungs following a normal expiration is the **functional residual capacity (FRC)**. The **inspiratory capacity (IC)** is the maximum amount of air inspired following a normal expiration. Figure 1 describes how the volumes and capacities described in the preceding paragraphs are related.

Figure 1. Inspiratory and expiratory lung volume

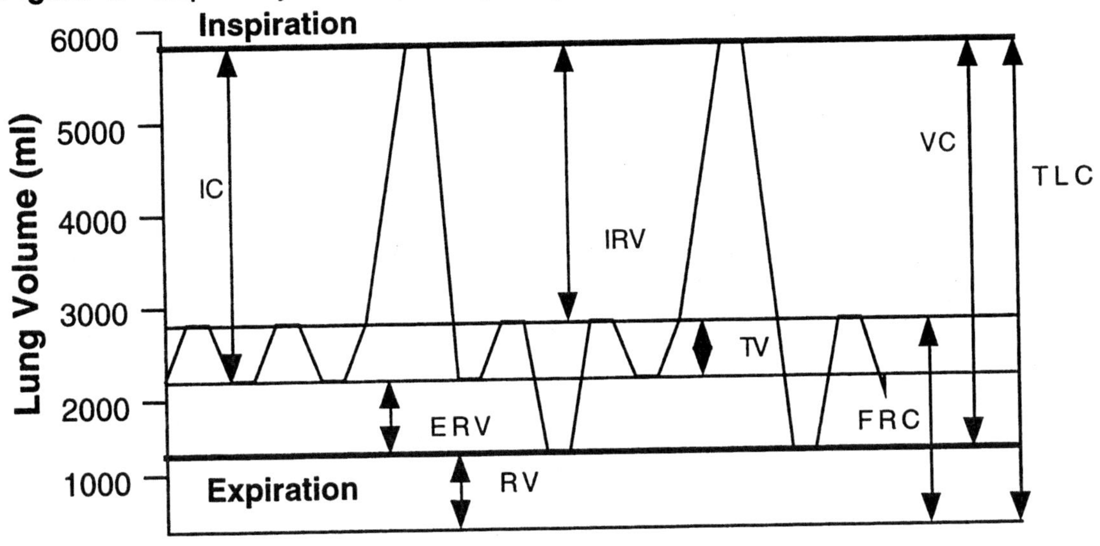

Figure 1 is an example of a spirogram reading. The spirogram is a recording of the volume of air inhaled and exhaled. Pen deflections (up and down) indicating an individual's breathing patterns are recorded on paper positioned around a vertical cylinder.

Helpful equations:

$$TLC = IRV + TV + ERV + RV = IC + FRC$$
$$VC = IRV + TV + ERV$$
$$FRC = ERV + RV$$
$$IC = IRV + TV$$

Clinicians use the spirogram to measure a patient's respiratory volumes (Figure 2). The spirogram reading allows one to determine the flow, or rate at which the air is exhaled, as well as the lung volumes. Analysis of the spirogram is one method a physician may use to diagnose a respiratory deficiency. Respiratory diseases fall into two classifications, obstructive and restrictive diseases.

Figure 2. Spirometer

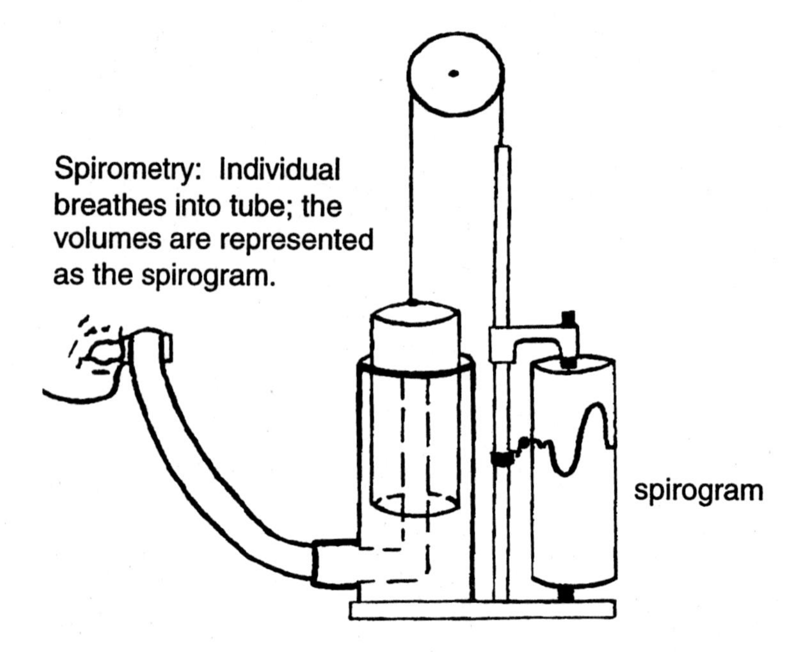

An individual with an **obstructive lung disease** exhibits decreased air flow due to increased airway resistance. Bronchoconstriction (narrowing of the airway passages) causes the increase in resistance. The most common obstructive disease in young adults is asthma. Especially pertinent to young adults is the issue of smoking. Smoke irritates the lungs, causing the airways to bronchoconstrict. If you smoke, you will develop an obstructive lung disease. Individuals suffering from obstructive lung diseases have a difficult time breathing (inspiring and expiring).

An individual suffering from a **restrictive lung disease** develops a decrease in the volume of air inside his or her lungs. The decreased volume may be due either to the chest's inability to expand or to a stiffness of the lungs. The occurrence of restrictive lung disease increases with age. Increased connective tissue within the lungs decreases the compliance (elasticity) and prevents the lungs from expanding fully. Problems in the chest wall may produce a decrease in lung volume, for example, diseases of the thoracic cage or diseases of the nerve supply to the respiratory muscles. In many ways the lung is just like a sponge. A wet sponge is soft and springs back after it has been compressed, but a dry sponge is hard and expands very little. The lungs from a person who has a restrictive lung disease would be like dry sponges. The lungs would hold much less air because they would not be able to expand to accommodate the air. Figure 3 illustrates the airway passages and the lungs in the nondiseased state. Narrowing of the airway passages causes an obstructive lung disease; decreased ability of the lungs to expand leads to a restrictive lung disease.

Figure 3. Airway passages leading into the lungs (nondiseased state)

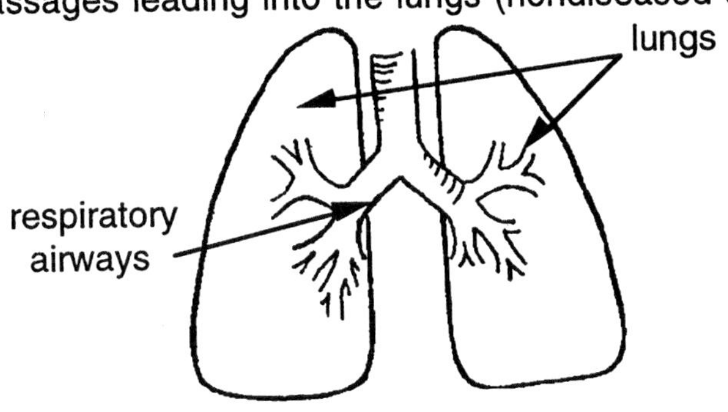

Part I: CHEST SIZE AND BODY POSITION

This activity is designed to determine the effect of body position on respiratory parameters, such as inspiratory capacity and vital capacity.

PROCEDURE

1. Students will pair up and alternately record data from one another.

2. The first student should exhale normally and then hold his or her breath while standing.

3. The student's partner will then measure the circumference of the chest cavity at the level of the xiphoid process. The xiphoid process is located at the end of the sternum, which joins the ribs. Follow the rib cage up to the middle of the chest. You should be able to feel the sternum; the xiphoid process is located at the tip of the sternum. Record data in box **A** of Table 1.

Figure 4. Location of the xiphoid process

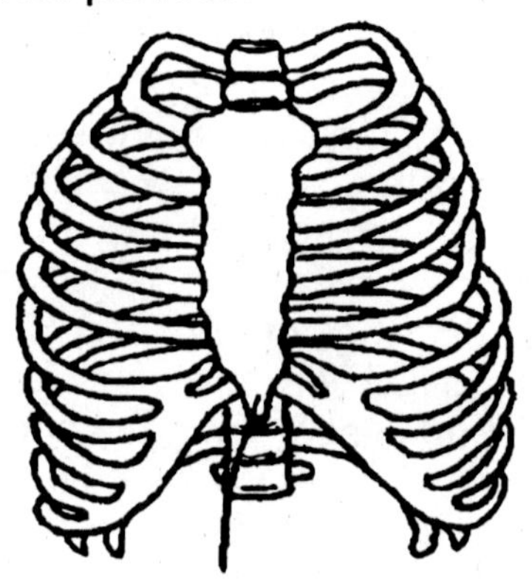

4. The first student should now inhale maximally and hold his or her breath.

5. The student's partner should then measure the circumference of the chest again and record it in box **B** of Table 1.

6. The first student should exhale maximally and hold his or her breath.

7. The students partner should then measure the circumference of the chest and record it in box **C** of Table 1.

8. The first student should now lie supine (on his or her back) on the floor.

9. Repeat the three measurements and record the data in boxes **A', B'** and **C'** of Table 1.

10. After the data has been collected exchange roles and repeat steps 2 through 9.

11. Now complete Table 2 to determine the inspiratory capacity and the vital capacity.

Table 1. Chest measurements (in cm)

A Standing (exhale normally)	B Standing (inhale maximally)	C Standing (exhale maximally)
A' Lying Down (exhale normally)	B' Lying Down (inhale maximally)	C' Lying Down (exhale maximally)

Table 2. Computations (in cm)

Inspiratory Capacity	Standing= B-A	Lying Down= B'-A'
Vital Capacity	Standing= B-C	Lying Down= B'-C'

ANALYSIS

1. What differences in chest size did you observe as positions changed?

2. How might these changes in size be explained?

Part II: HEIGHT VERSUS VITAL CAPACITY

This activity is designed to determine if there is a relationship between an individual's height and his or her vital capacity.

PROCEDURE

1. Complete Table 3 by recording the height (cm) and vital capacity of all the students in the class.

2. Plot the data on Graph 1, representing the height on the *x*-axis and the vital capacity on the *y*-axis.

3. Once all the points are plotted, construct a regression line.

Table 3. Height versus Vital Capacity

NAME	HEIGHT (cm)	VITAL CAPACITY	NAME	HEIGHT (cm)	VITAL CAPACITY

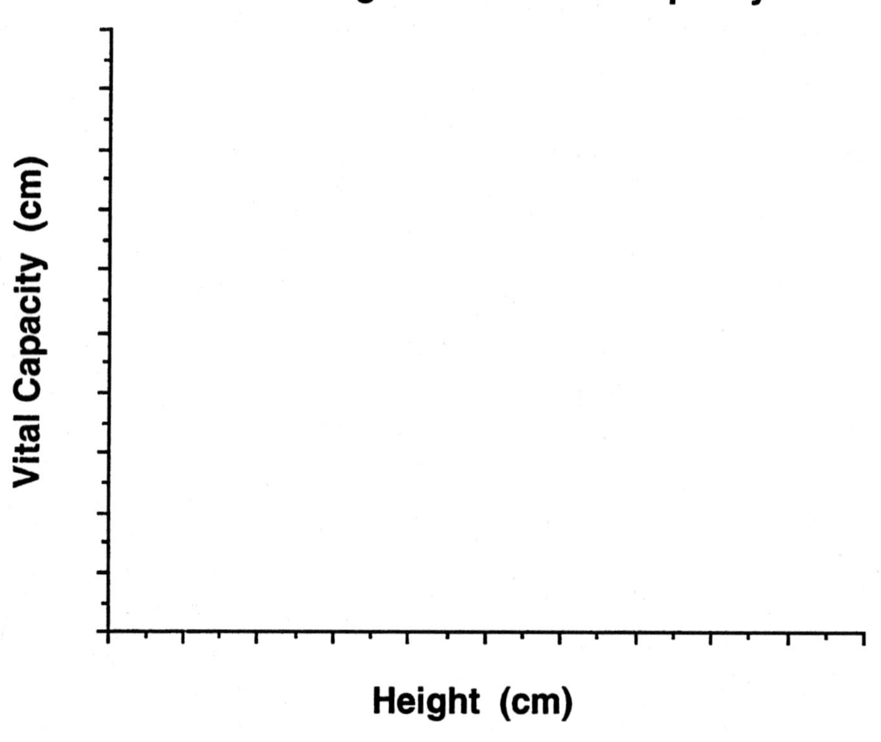

ANALYSIS

1. What relationship did you find between height and vital capacity?

2. What might be the explanation for this relationship?

Part III: THE EFFECTS OF DISEASE ON RESPIRATION

Chronic Obstructive Respiratory Conditions

This activity is designed to determine the effect of a simulated obstructive disease on inspiratory and vital capacities.

PROCEDURE

1. Repeat steps 2–11 in Part I, but place a clip on the student's nose and have him or her breathe through a straw while performing the activity. The straw will narrow the airway and therefore simulate an obstructive disease.

2. Record your measurements in Table 4a and perform the calculations in Table 4b.

Restrictive Respiratory Conditions

This activity is designed to determine the effect of a simulated restrictive disease on inspiratory and vital capacities.

PROCEDURE

1. Repeat steps 2–11 in Part I, but fasten a belt around the chest cavity. The belt will restrict the expansion of the chest and therefore simulate a restrictive disease.

2. Record your measurements in Table 4 and perform the calculations in Table 4b.

Table 4a. Chest circumference from simulated restrictive and obstructive diseases

CONDITION:	Exhale Normally	Inhale Maximally	Exhale Maximally
Normal			
Obstructive	(A) =	(B) =	(C) =
Restrictive	(A')=	(B')=	(C')=

TABLE 4b. Computations (in cm)

Inspiratory Capacity	Chronic Obs. = B-A	Restrictive = B'-A'
Vital Capacity	Chronic Obs. = B-C	Restrictive = B'-C'

ANALYSIS

1. How do the conditions imitating obstructive respiratory diseases affect chest size compared with normal conditions?

2. How does the chest size in restrictive respiratory diseases differ when compared with normal?

Part IV: ADDITIONAL INVESTIGATIONS

Design your own investigation by choosing one idea from the following list. Your investigation should be based on the mean obtained from the measurements of at least five individuals. You can examine any aspect of chest size (inspiratory capacity, vital capacity, etc.) studied in this lesson.

1. **Prone vs. Supine** (class example)
   Is there a difference in chest size while lying on your back (supine) compared to lying on your stomach (prone)?

2. **Athlete vs. Nonathlete**
   How much of a difference, if any, is there between the respiratory responses of those who play sports and those who do not?

3. **Athlete vs. Athlete**
   Does one sport hold any advantage over another in conditioning for respiratory responses?

4. **Male vs. Female**
   Which gender has the greater vital capacity? (Pick individuals of approximately the same height for this study).

LABORATORY EXERCISE 5

ENDOCRINE AND REPRODUCTIVE PHYSIOLOGY

BACKGROUND AND KEY TERMS

The organs of the body communicate with each other in order to coordinate their activities. People use telephones and the postal service; our bodies use the nervous and endocrine systems. The endocrine system is made up of hormones, their releasing organs, and their target organs. **Hormones** are chemicals produced by specific tissues in the body and released into the bloodstream. The body's nervous system is comparable to the telephone system because it sends fast, direct messages. The endocrine system is comparable to our mail system because the delivery of the message is slower. Like bulk mail, the message is more diffuse (reaches a greater area) and affects many organs. A hormone travels through the body via the blood but affects only the cells with receptors for that specific hormone. Hormones are a slower method of communication, but their effects are longer lasting.

The **hypothalamus** is a small, penny-sized portion of the brain. The hypothalamus acts as the command center for the endocrine system. It can act either as a regulator of the anterior pituitary gland or as an endocrine organ that releases oxytocin and antidiuretic hormone (**ADH**); ADH is also called vasopressin. For this experiment, we will focus on the hypothalamus as a regulator of the pituitary gland. The hypothalamus releases hormones such as thyroid releasing hormone (**TRH**) and gonadotropin releasing hormone (**GnRH**), which travel through a special blood vessel system connected to the pituitary gland. When the hormones (TRH and GnRH) reach the pituitary gland, they stimulate the pituitary gland to release hormones that then travel to their target organs. For example, the hypothalamus releases TRH, which travels to the pituitary gland and stimulates the release of thyroid stimulating hormone (**TSH**); TSH travels to the thyroid (the target organ) and stimulates the release of thyroid hormone.

The endocrine system keeps itself in check through feedback systems. The **negative feedback** system is similar to a thermostat. An increase in temperature stimulates the air conditioner to blow out cold air. When the air conditioner forces out so much cold air that the room is below the desired temperature, the thermostat detects this condition and turns off the air conditioner. In order to keep the room temperature fairly constant, the thermostat assesses the situation and turns the air conditioner on and off accordingly. The endocrine system regulates the levels of hormone found in the body's bloodstream in the same manner. In negative feedback the hormone released from an organ "feeds back" or communicates with the organ to stop the release of any more hormone because a sufficient amount of hormone has already been released. It is important to remember that negative feedback also inhibits organs higher up in the chain of command. For example, an excess of cortisol released from the adrenal glands will inhibit the further production and secretion of cortisol as well as inhibit the pituitary gland from releasing adrenocorticotropin hormone (**ACTH**), which stimulates the adrenal glands to release cortisol. Excess cortisol will also inhibit the release of corticotropin releasing hormone (**CRH**) from the hypothalamus, which stimulates the release of ACTH from the pituitary gland. In other words, the gland has released enough hormone to fulfill its function; this is sensed by the body and production of the hormone ceases. There is also **positive feedback**, in which the end product further stimulates the releasing organ. This form of feedback is less common.

For the purposes of this experiment, we will look at only one of the hypothalamic releasing hormones: thyroid releasing hormone (TRH). This hormone travels to the anterior pituitary gland via the bloodstream to stimulate production of thyroid stimulating hormone (TSH). It is important to know that the hypothalamus secretes a releasing hormone to regulate each of the hormones released from the anterior pituitary gland. In this way, the hypothalamus is like a command center. If the hypothalamus is not stimulated, hypothalamic releasing hormones will not stimulate the anterior pituitary to release its hormones.

The glands of the body increase in size much like muscle when they are excessively stimulated. This condition is called **hypertrophy**. For example, the lymph nodes become enlarged when we are sick in order to release more lymphocytes to fight the infection; lymphocytes are cells that protect the body against foreign particles (infections). In a corresponding manner, if a gland is continuously inhibited, it will shrink in size or **atrophy**. For example, increased levels of cortisol in the blood over an extended period of time will inhibit the adrenal glands (through negative feedback), causing the glands to decrease in size.

In addition to TSH and ACTH, the anterior pituitary gland also releases the hormones luteinizing hormone (**LH**), follicle stimulating hormone (**FSH**), growth hormone (**GH**), and prolactin. Each of these hormones is released into the bloodstream to affect a specific target organ.

Thyroid stimulating hormone (TSH) travels to the thyroid gland to stimulate the production and release of **thyroid hormone**. Thyroid hormone influences the growth rate of many body tissues and is necessary for proper central nervous system development. Its main function is to increase a person's basal metabolic rate (BMR) and to increase heat production. **Hyperthyroidism** is the excessive production of thyroid hormone. The most common cause of hyperthyroidism is Grave's disease; the symptoms include increased BMR, a constant feeling of warmth, nervousness, and goiter. Goiter is the term used to describe an enlarged thyroid gland. The enlargement of the thyroid may be due to a defect anywhere in the pathway from the hypothalamus to the pituitary gland to the thyroid gland. Decreased levels of thyroid hormone, or hypothyroidism, may also occur. The symptoms of hypothyroidism are low BMR, decreased appetite, abnormal central nervous system development, and intolerance to cold.

Adrenocorticotropin hormone (ACTH) is released in response to corticotropin releasing hormone (CRH). The release of CRH is regulated by negative feedback, circadian rhythms, and stress. ACTH released from the anterior pituitary gland stimulates the adrenal glands to secrete **cortisol**. Under normal conditions, excess cortisol in the bloodstream will inhibit this pathway by negatively feeding back to the hypothalamus and the anterior pituitary, as visually represented in Figure 1. By using a negative feedback system, the body produces only the amount of hormone it needs without wasting its resources. Cortisol promotes the breakdown of proteins and fats and helps the body adapt to stress. Cortisol also acts as an immunosuppressive and anti-inflammatory drug. If cortisol is administered in large doses, its immunosuppressive properties will cause the organs of the immune system to shrink. **Cushing's syndrome** is a condition of excess secretion of cortisol, or hypercortisolism. The symptoms of Cushing's syndrome include personality changes, hypertension (high blood pressure), osteoporosis (weakening of bones due to loss of calcium), and weight loss. The protein degradation caused by cortisol leads to a "wasting" effect if an excess level of cortisol remains in the body. Hyposecretion of cortisol is characterized by symptoms such as defective metabolism, mental confusion, and a decreased ability to adapt to stress.

Figure 1. Negative feedback control

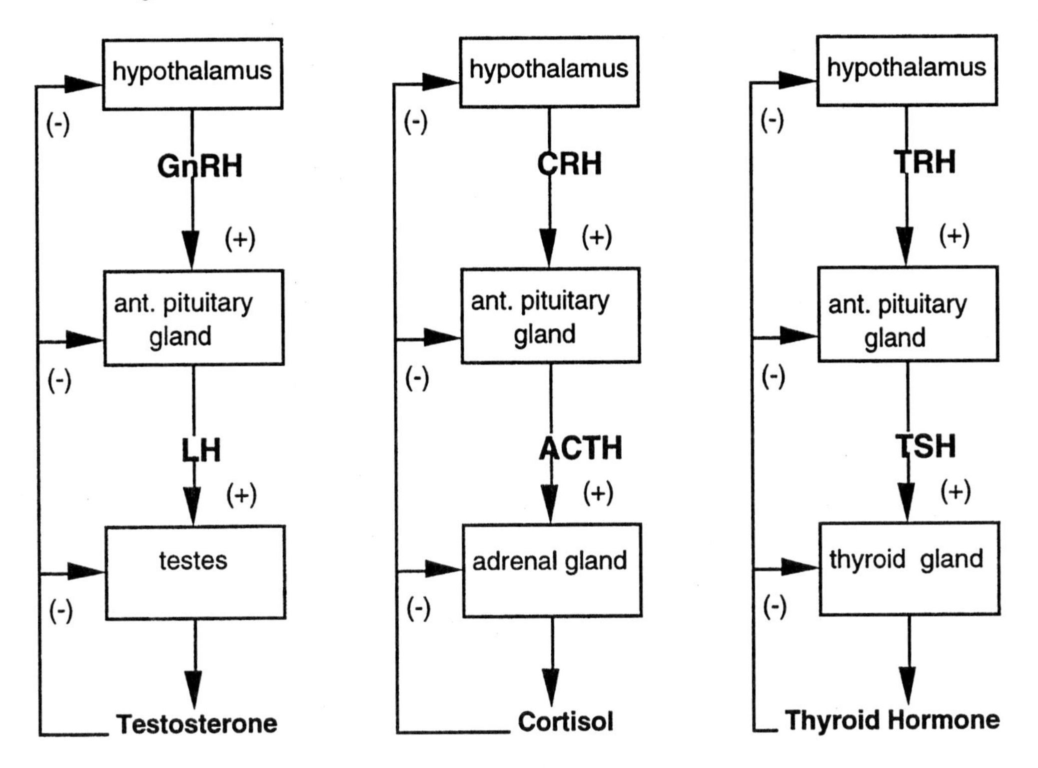

The anterior pituitary gland releases luteinizing hormone (LH), which travels to the **Leydig cells** of the testes in the male. Leydig cells are found in the connective tissue between the **seminiferous tubules**. Upon stimulation with LH, the Leydig cell releases **testosterone**. Testosterone is responsible for the male sex drive and secondary sex characteristics, such as increased body hair and deep voice. Negative effects of testosterone are male pattern baldness and increased secretion of the sebaceous glands, which can lead to acne. Decreased amounts of testosterone in the body primarily affect the sexual organs. If testosterone levels are low, males will not develop normally and will have sperm counts too low to fertilize an egg. The condition of excess levels of testosterone is rare but causes in premature sexual development.

Figure 2. Organs of the male reproductive tract

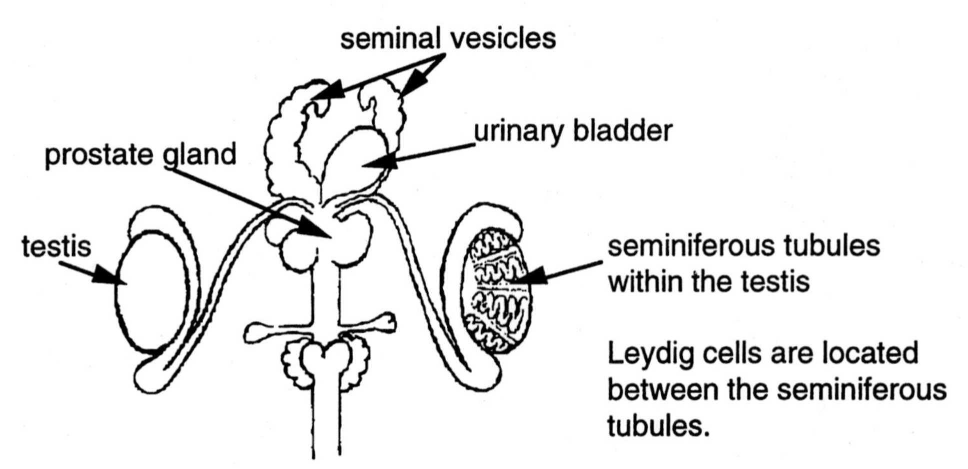

In the female, LH causes the **follicle** (developing egg) in the ovary to secrete **estrogen**. The estrogen participates in either a positive or negative feedback loop. LH causes ovulation (release of egg from the ovary) in the female; there is a large rise in levels of LH just prior to ovulation due to positive feedback from estrogen; the secretion of estrogen from the follicle stimulates the release of even more LH. Estrogen causes the development of female secondary sexual characteristics and sustains the female reproductive tract. A woman who lacks ovaries (and therefore follicles) will not produce estrogen. However, the pituitary gland will secrete excess LH because there is no mechanism to turn off the release of LH from the pituitary. Estrogen regulates the release of LH by negative feedback during the pre- and postovulatory phases. Excess levels of estrogen cause early sexual development in the female.

DETERMINATION OF UNKNOWN HORMONES

This exercise is designed to determine the identity of an unknown hormone by observing the effect it (the hormone) had on the organs of the male rat.

PROCEDURE

The data for this laboratory were compiled from seven sets of male laboratory rats, two rats per set; one set was the control group and the remaining six were experimental groups. The rats were all male in order to study the relationship between the reproductive and endocrine systems. The female reproductive system is more difficult to study than the male reproductive system because it is continuously cycling. In each set of rats there was an "intact" rat and a "castrate rat." The castration involved removal of the testes to eliminate testosterone production. The two rats (normal and castrate) of each group were treated alike in all other ways (food, water, etc.). All rats, except for those in the control group were injected with a hormone on a daily basis for 2 weeks. Autopsies were performed on the animals at that time.

The group of medical students performing this exercise were very disorganized and rushed through the work, making errors in labeling the bottles of hormone. The students obtained the following results for organ weights after the autopsies were performed. In this short period of time, the students noted amazing changes in the size of certain organs when compared the experimental group of rats with the control group. Using the flowchart (Figure 1), Table 1, and the autopsy data, match the unknown rat groups with their respective hormones. The bottles on the refrigerator shelf were ACTH, cortisol, LH, TRH, testosterone, and TSH.

Figure 3 represents the organs of the rats used in the experiment. The pituitary is not drawn to scale; it was drawn larger than actual size.

Figure 3. Graphic representation of organs studied in the autopsy

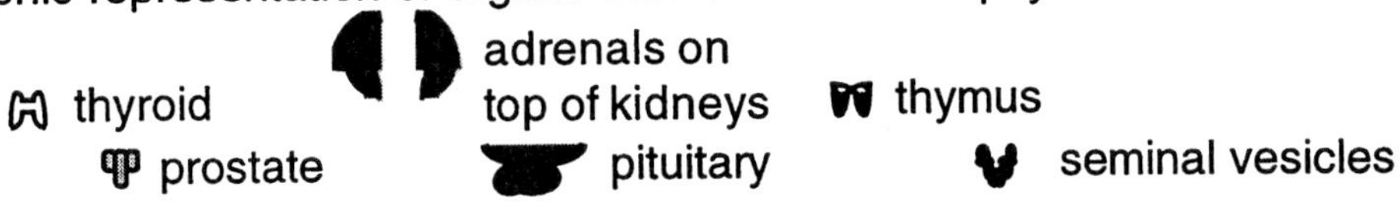

thyroid
prostate
adrenals on top of kidneys
pituitary
thymus
seminal vesicles

Table 1. Effects of excess hormones on the following glands; + represents a positive effect, i.e., hypertrophy; - represents a negative effect, i.e., atrophy; NC refers to no change in size.

					Testosterone		LH	
	TRH	TSH	ACTH	Cortisol	Intact	Castrate	Intact	Castrate
Pituitary gland	+	-						
Thyroid gland	+	+						
Adrenal glands			+	-				
Thymus gland			-	-				
Testes							+	NC
Prostate					+	+	+	NC
Sem. vesicles					+	+	+	NC
Body weight	-	-	-	-	+	+		
Organ size	-	-						

ANALYSIS

1. What was hormone 1? Explain your answer.

2. What was hormone 2? Explain your answer.

3. What was hormone 3? Explain your answer.

4. What was hormone 4? Explain your answer.

5. What was hormone 5? Explain your answer.

6. What was hormone 6? Explain your answer.

This is your set of control rats; the data are the results of the autopsy.

Control (intact)

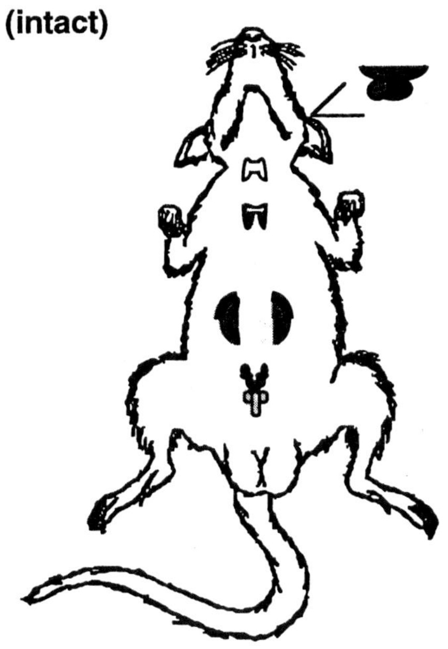

Pituitary: 12.9 mg

Thyroid: 250 mg

Thymus: 475 mg

Adrenals: 40 mg

Seminal vesicles: 500 mg

Prostate: 425 mg

Testes: 3200 mg

Body weight: 300 g

Control

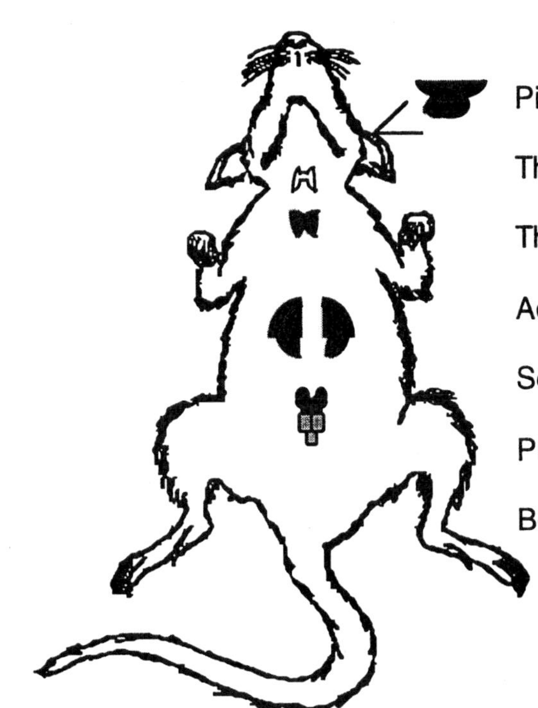

Pituitary: 12.9

Thyroid: 250 mg

Thymus: 480 mg

Adrenals: 40 mg

Seminal vesicles: 450 mg

Prostate: 387 mg

Body weight: 270 g

Determine the identity of hormone 1 using the following data from the autopsy, Table 1, and the negative feedback flowchart (Figure 1).

Hormone 1
(intact)

Pituitary: 12.8 mg

Thyroid: 245 mg

Thymus: 150 mg

Adrenals: 100 mg

Seminal vesicles: 490 mg

Prostate: 430 mg

Testes: 3000 mg

Body weight: 200 g

Hormone 1

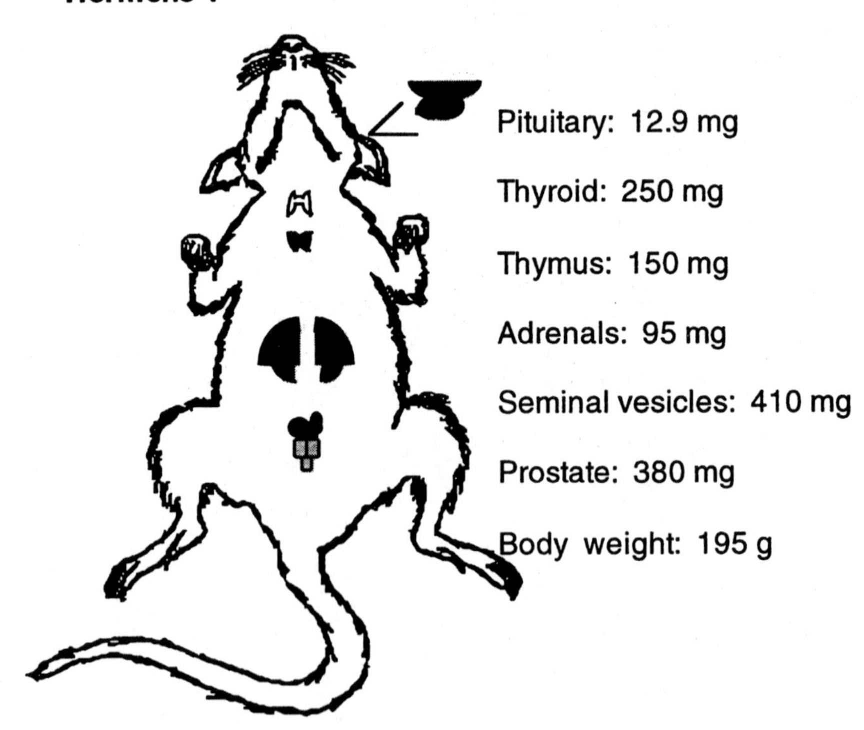

Pituitary: 12.9 mg

Thyroid: 250 mg

Thymus: 150 mg

Adrenals: 95 mg

Seminal vesicles: 410 mg

Prostate: 380 mg

Body weight: 195 g

Determine the identity of hormone 2 using the following data from the autopsy, Table 1, and the negative feedback flowchart.

Hormone 2
(intact)

Pituitary: 13.0 mg

Thyroid: 250 mg

Thymus: 480 mg

Adrenals: 40 mg

Seminal vesicles: 900 mg

Prostate: 800 mg

Testes: 5700 mg

Body weight: 310 g

Hormone 2

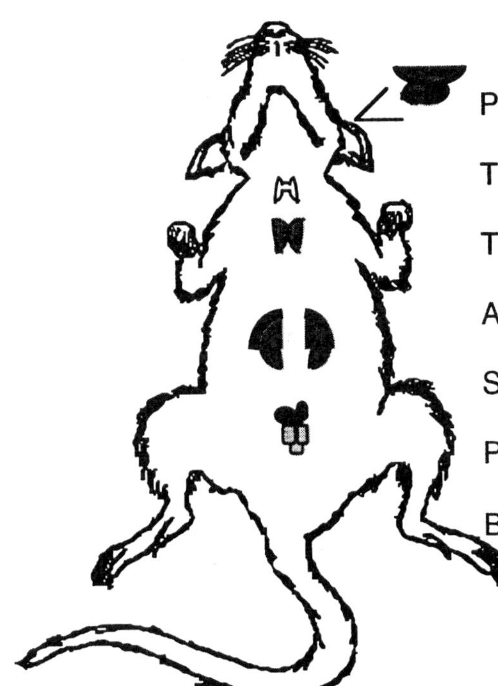

Pituitary: 13 mg

Thyroid: 250 mg

Thymus: 480 mg

Adrenals: 42 mg

Seminal vesicles: 412 mg

Prostate: 375 mg

Body weight: 275 g

Determine the identity of hormone 3 using the following data from the autopsy, Table 1, and the negative feedback flowchart.

Hormone 3
(intact)

Pituitary: 13.2 mg

Thyroid: 252 mg

Thymus: 470 mg

Adrenals: 38 mg

Seminal vesicles: 1400 mg

Prostate: 900 mg

Testes: 3000 mg

Body weight: 400 g

hormone 3

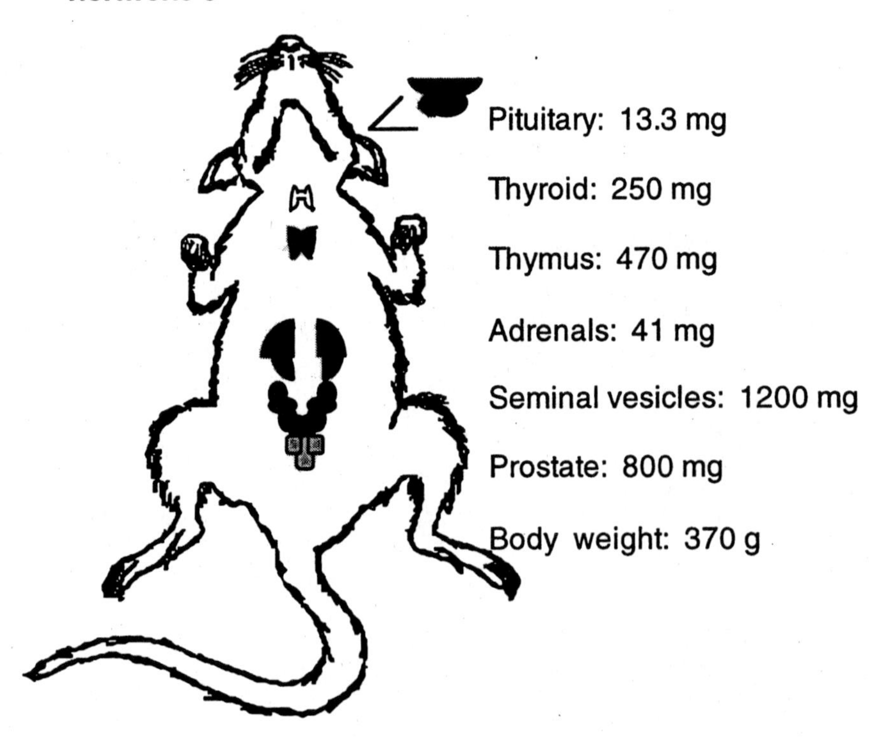

Determine the identity of hormone 4 using the following data from the autopsy, Table 1, and the negative feedback flowchart.

Hormone 4
(intact)

Pituitary: 25 mg

Thyroid: 490 mg

Thymus: 462 mg

Adrenals: 39 mg

Seminal vesicles: 480 mg

Prostate: 400 mg

Testes: 1650 mg

Body weight: 160 g

Hormone 4
(castrate)

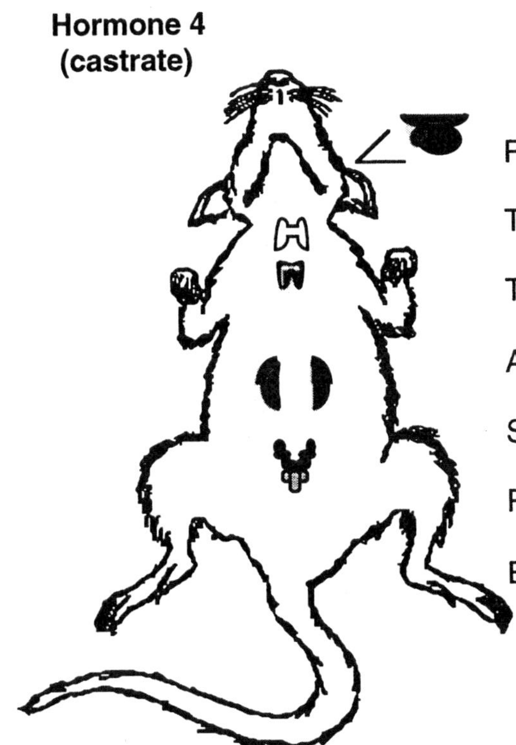

Pituitary: 25.7 mg

Thyroid: 495 mg

Thymus: 460 mg

Adrenals: 38 mg

Seminal vesicles: 450 mg

Prostate: 375 mg

Body weight: 144 g

Determine the identity of hormone 5 using the following data from the autopsy, Table 1, and the negative feedback flowchart.

Hormone 5
(intact)

Pituitary: 13 mg

Thyroid: 245 mg

Thymus: 250 mg

Adrenals: 35 mg

Seminal vesicles: 475 mg

Prostate: 410 mg

Testes: 3200 mg

Body weight: 150 g

Hormone 5

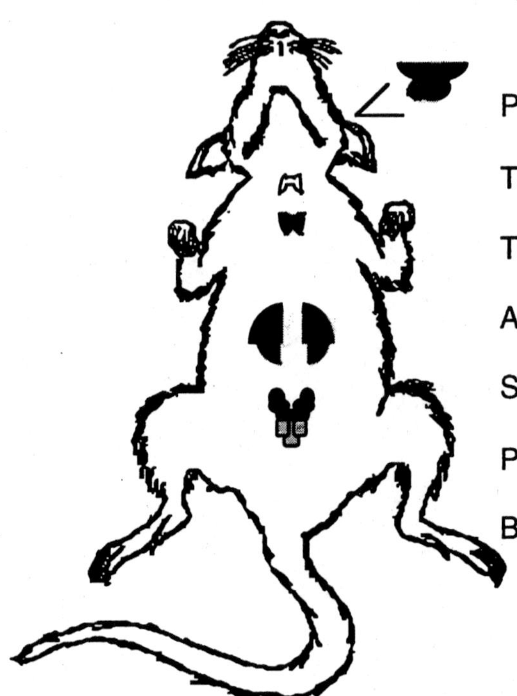

Pituitary: 12.9 mg

Thyroid: 247 mg

Thymus: 240 mg

Adrenals: 35 mg

Seminal vesicles: 440 mg

Prostate: 380 mg

Body weight: 135 g

Determine the identity of hormone 6 using the following data from the autopsy, Table 1, and the negative feedback flowchart.

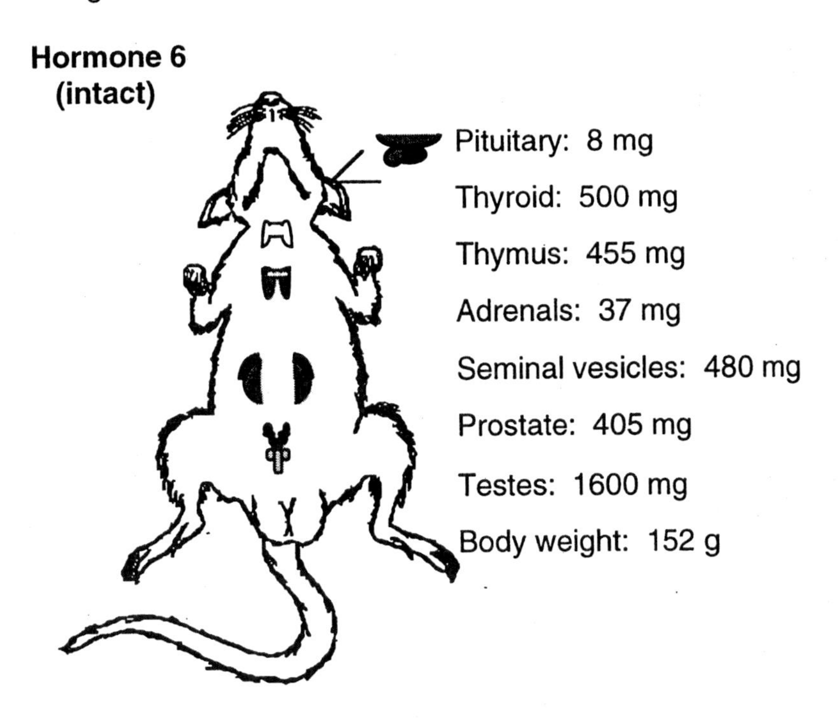

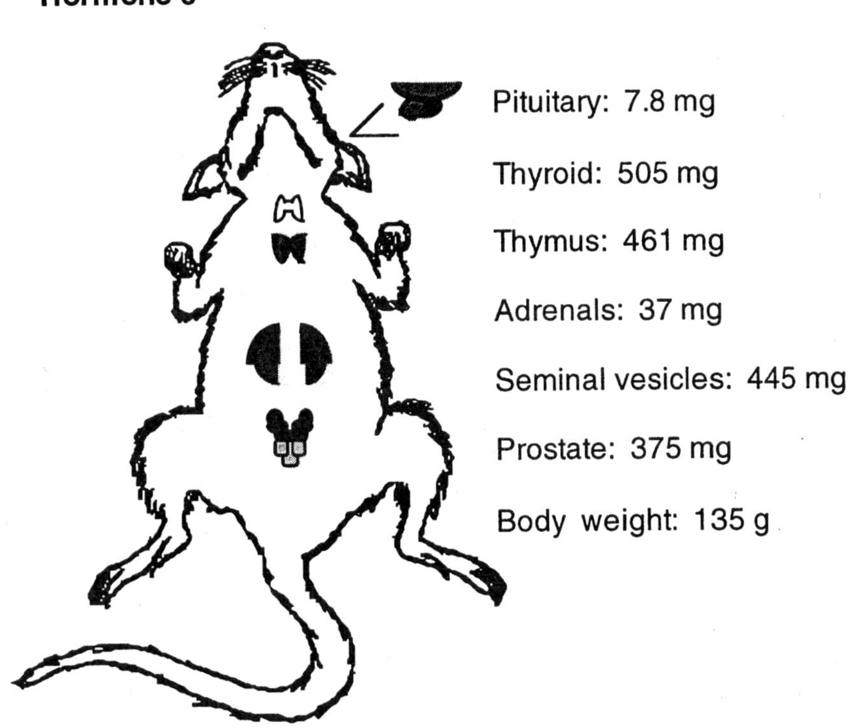

LABORATORY EXERCISE 6

THE EFFECT OF TEMPERATURE ON BLOOD FLOW

BACKGROUND AND KEY TERMS

It is widely known that heat and cold are applied to injuries. Do you know why?

The body has many mechanisms for keeping itself in equilibrium. When the body experiences change (blood pressure, core temperature, or heart rate), a compensating reaction occurs to return it to its normal state. Changes in temperature are part of this equilibrium mechanism. Our bodies control temperature by changing blood volume and the rate of blood flow. Blood flow is primarily controlled through vasoconstriction (narrowing of blood vessels) mediated by the sympathetic nervous system. The sympathetic nervous system is a division of the autonomic nervous system, which controls involuntary activities such as heart rate, body temperature, arterial blood pressure, and gastrointestinal activity. When stimulated by the sympathetic nerves, the smooth muscle in the blood vessel wall contracts, constricting the vessel. This is known as **vasoconstriction**. In general, vasoconstriction is used by the body to conserve heat, and **vasodilation** (widening of blood vessels) is used to release heat. Interestingly, both heat and cold provide relief from pain.

Cold is often used in the acute (immediately after an injury) phase of an injury. After a cold pack (or ice wrapped in cloth) is applied to the area for about 5 minutes, vasodilation is evidence by reddening of the local area. Hyperemia (reddening) is due to the increased flow of blood. At this point, even though there is a decrease in local skin temperature, the body is still able to bring in enough warm blood to "dilute" the effects of the cold treatment. Pallor (whiteness) is seen after 10–15 minutes. At this point, less blood is flowing to the area due to vasoconstriction, which the body uses to prevent heat loss. The vasoconstriction is very helpful in the case of injuries, because with less blood flow there will be less fluid leakage (edema or swelling) from the ruptured vessels. Also, cold effectively relieves pain associated with muscle spasms. However, with prolonged exposure to cold, a second period of vasodilation is seen. In an extreme case of cold exposure (frostbite) there is actually coagulation of the blood; the skin appears blue because no newly oxygenated blood is flowing into the frostbitten area.

Chemical reactions are temperature sensitive. The speed of a chemical reaction decreases by half for every drop in temperature of 10 degrees Celsius (^{o}C). Similarly, nerve conduction decreases by 1.84 m/s per degree Celsius between 36 and 23 ^{o}C. With prolonged application of a cold pack, vasodilation is seen after the initial vasoconstriction. The tissue reaches such a low temperature that nerve conduction stops. Passive dilation occurs because the nerves are no longer stimulating the vessels to constrict. The decrease in temperature lowers the metabolic rate in the region where the cold pack is applied. Venous blood returning from the area of treatment contains more oxygen because the tissue is not utilizing as much of it for metabolic processes.

Application of heat is used in the chronic phase of injury. Heat causes a local increase in metabolism. The increased metabolites (byproducts of metabolism) stimulate the release of histamine, a vasodilator substance. In addition to the active vasodilation caused by histamine, passive dilation also occurs due to the removal of sympathetic vasoconstrictor tone. Vasodilation disseminates the energy, allowing cooler blood to enter and balance the heat energy being absorbed. The first indication that the heat is being applied at a dangerous level is the appearance of white and red blotches, referred to as *mottling.* The vessels have been maximally dilated for so long that spasm occurs in the smooth muscle of the blood vessel walls, which results in vasoconstriction, represented by the patchy white areas. The hypothalamus regulates the body's core temperature because it is sensitive to changes in the temperature of the blood flowing to it. The localized heating causes the brain to withdraw sympathetic vasoconstrictor tone, which leads to passive dilation. If the heat is transferred faster than it is dissipated, then an increase in temperature will occur. Maximum vasodilation occurs 20–30 minutes after application of heat. At this time, it is possible to experience a slight decrease in skin temperature because of the high volume of blood flowing to the area to disseminate the heat.

Indirect heating is the process in which an area of the body that is *not* being heated directly exhibits vasodilation due to the influence of the brain. A similar phenomenon exists with the application of cold. Vasoconstriction occurs in areas other than the local area that is being treated. In the clinical setting it is important to determine if the patient has high blood pressure before applying a cold treatment because the reflex (indirect) vasoconstriction can lead to a large rise in blood pressure.

Skin neurons that are sensitive to touch vary in number and distribution throughout the body surface. One way to determine how close they are to each other is to perform a tactile (touch) test, which examines a subject's ability to discriminate between two points touching his or her skin. Temperature, especially a low temperature, has an effect on two-point discrimination. This is tested by having the experimenter touch the student's skin with a device known as a two-point stimulator (or a two-point threshold device). It consists of two pins that are individually fixed to a ruler. The points can be adjusted to vary the distance between them. The subject should close his or her eyes and report how many pins he or she feels as the experimenter touches the skin lightly with the points of the device. Initially the pins are placed closest together. The distance between the points is gradually increased until the subject is able to feel two points. This distance between the pins represents the space between the tactile neurons.

Part I: THE EFFECTS OF DIRECT AND INDIRECT HEAT ON BLOOD FLOW

This exercise is designed to examine the effect of indirect and direct heat on blood flow.

PROCEDURE

1. Fill a bucket or small pail about one-third full of warm tap water at a temperature between 42 and 45 ^{o}C. Use an alcohol thermometer to monitor the temperature.

2. Using a surface thermometer, record the initial skin surface temperature of the back of the left hand. Record this value in Table 1.

3. Place your right hand in the bucket of water. Keep the hand in the bucket for the next 20 minutes. Maintain the 42–45 ^{o}C temperature by adding more warm water as needed.

4. Record temperature readings on the left hand once each minute during the 20-minute time period.

5. After 20 minutes, remove your right hand from the water and allow time for both hands to return to their initial temperatures.

6. Repeat steps 1–5 but this time cover the left hand with a towel or blanket. For best results, make sure the legs and arms are also covered (long pants, sweatshirt). Uncover only a small portion of the left hand when taking the temperature.

7. Using the data from Table 1 obtained from the measurements of the bare and covered left hand, make two line graphs on Graph 1. Use a solid line to represent the bare hand and a dashed line to represent the covered hand.

Table 1. Temperature change in the bare and covered left hand

Time (min)	Left Hand (bare)	Left Hand (covered)
0		
1		
2		
3		
4		
5		
6		
7		
8		
9		
10		
11		
12		
13		
14		
15		
16		
17		
18		
19		
20		

GRAPH 1. Left-Hand Temperature (Bare and Covered) vs.Time

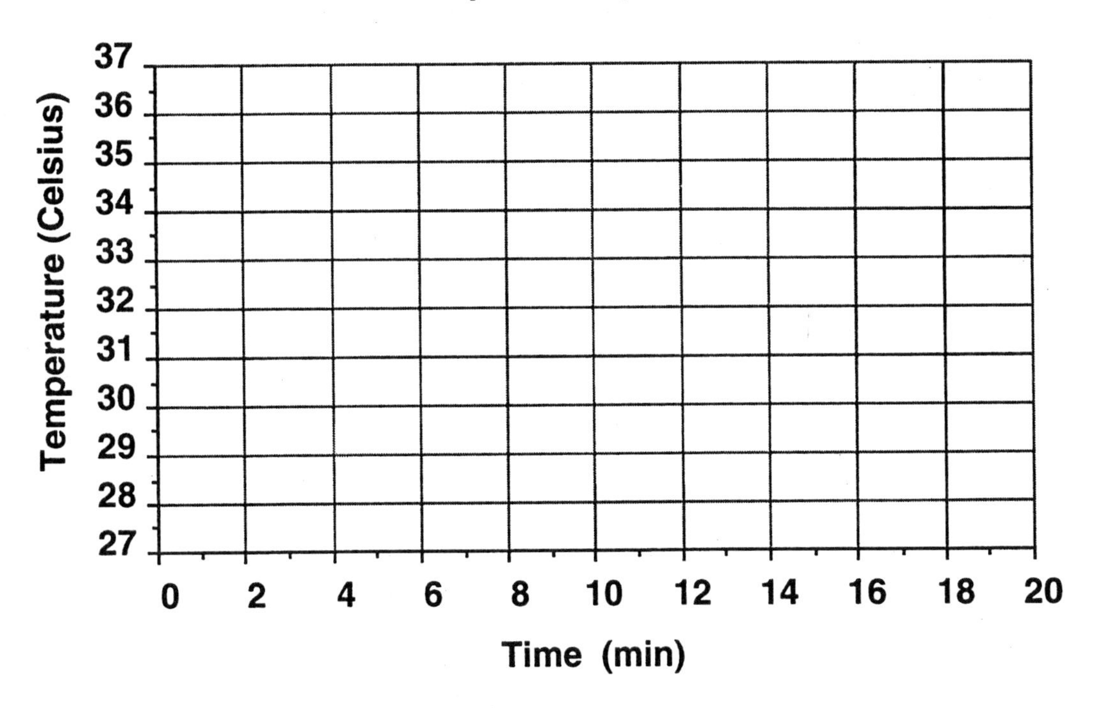

ANALYSIS OF PART I

1. Which hand (bare or covered) had the greater change in temperature?

2. Why was the hand covered? Discuss the factors that would cause the observed change in temperature of the hand.

Part II: THE EFFECTS OF COLD ON BLOOD FLOW

This exercise is designed to examine the effects of cold on blood flow.

1. Divide the class into a group of subjects and a group of data recorders.

2. Using the two-point threshold device, measure the two-point discrimination abilities of each of the subjects. The open right hand of the subject should lie on the table with the palm up. The eyes of the subject are to be closed. Lightly touch the center of the palm with the two point discrimination device while the subject responds with the words "one" or "two", describing what was felt. Begin with the pins set for no space between them and gradually increase the distance by 1 mm each time until the subject responds with a "two". Record measurements in Table 2.

3. While the subject's eyes are still closed, give a coin discrimination test. Place four coins (penny, nickel, dime, and quarter) in random order one at a time on the subject's palm. Have the student guess which coin it is. Before administering the test, permit one practice trial in which the subject is allowed to guess the coins and then is given the correct answer. The practice trial will acquaint the subject with the feeling of the coins. Record the number of correct responses obtained in the real test in Table 2.

4. Pair up an observer with a subject and record a description of the color of *both* hands of the subject in Table 3.

5. Have the subject grasp a handful of crushed ice with the right hand. Have him or her hold the hand over the bucket so the ice doesn't drip on the floor. After 2, 5, 10, and 15 minutes, record the appearance and surface temperature of both hands in Table 3. Replace the ice after each recording point or if the ice melts before the end of the test. During the experiment, report any sensations that the subject reports feeling.

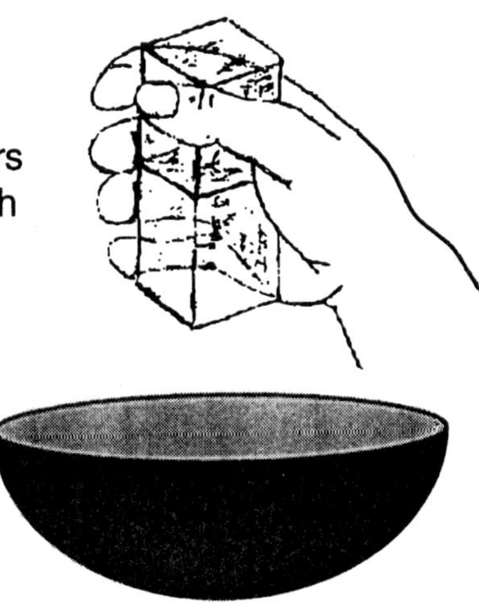

6. After 20 minutes, repeat the two-point and coin discrimination tests. Record the answers in Table 2. Have the student apply pressure on the cold hand with his or her warm hand. Note the changes. Take a final surface temperature and record the value. Graph the surface temperature of both hands on Graph 2, using a solid line for the cold hand and a dashed line for the warm hand.

Table 2. Comparison of skin neuron sensitivity with discrimination tests before and after the application of ice

	Before Ice	After Ice
Two-point discrimination (mm)		
Coin accuracy (number correct)		

Table 3. Temperature changes and observations of right and left hands during the application of ice

Time	Temperature		Observations	
	L. Hand	R. Hand	Left Hand	Right Hand
0 min				
2 min				
5 min				
10 min				
15 min				

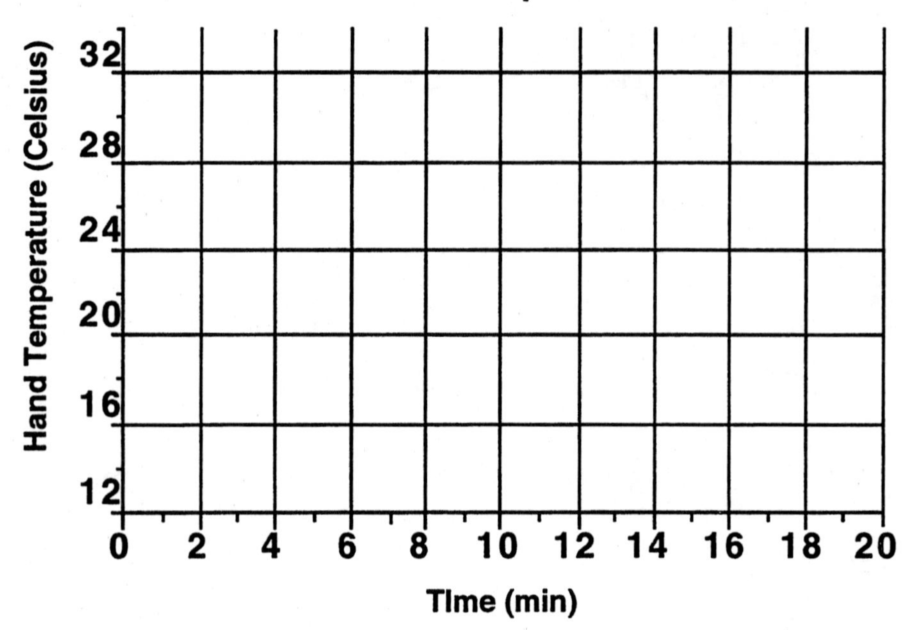

ANALYSIS OF PART II

1. How did the appearance of the hands change over time?

2. What changes occurred in the warm hand? Why?

3. What do the temperature changes show?

4. Why does the hand go numb?

5. What differences are noted in two-point discrimination before and after the experiment?

6. What occurs when pressure is applied to the cold hand?

LABORATORY EXERCISE 7

REFLEXES

BACKGROUND AND KEY TERMS

A reflex is defined as an automatic response to a stimulus. It is carried out by a simple neuronal pathway (see Figure 1) that consists of a receptor, afferent tract, efferent tract, and an effector organ. The afferent tract consists of nerves that carry impulses to the central nervous system (CNS), and the efferent tract carries signals away from the CNS. Our bodies use reflexes every day without our realizing it.

Figure 1. Neuronal pathway of a reflex

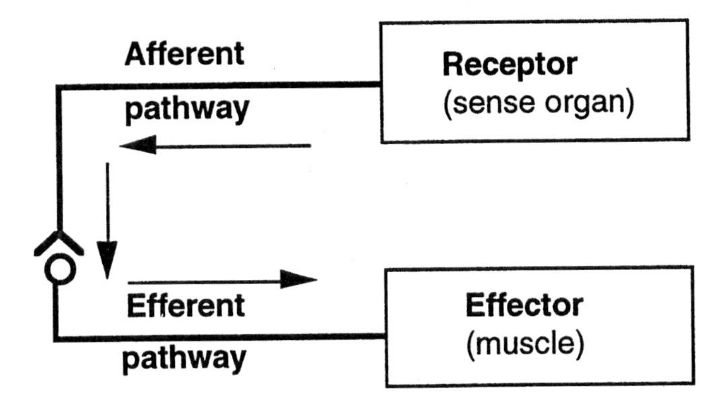

Daily activities such as blinking, swallowing, and walking are all reflexes. The simplest type of reflex is the **monosynaptic reflex.** In this type of reflex a receptor senses the change in conditions, and an afferent neuron transmits the signal to the spinal cord; the afferent neuron synapses (communicates) directly with the efferent neuron, which transmits a message to the effector organ; the effector organ then carries out the appropriate action (described in the message carried down the efferent neuron).

The **muscle stretch reflex**, illustrated in Figure 2, is a monosynaptic reflex; the "receptor" is the muscle **spindle**, a small organ that senses lengthening or stretching of the muscle fibers, and the "effector organ" is the muscle that has been stretched. When a muscle stretches, the spindles located in the muscle sense the stretch and cause the muscle to contract (shorten). The **knee jerk reflex** operates by this mechanism; tapping the patellar tendon with a reflex hammer causes a slight stretch of the quadriceps muscle. The muscle spindles in the quadriceps sense the stretch and cause the quadriceps to contract, which makes the lower leg kick outward.

Figure 2. Pathway of the muscle stretch reflex, an example of a monosynaptic reflex

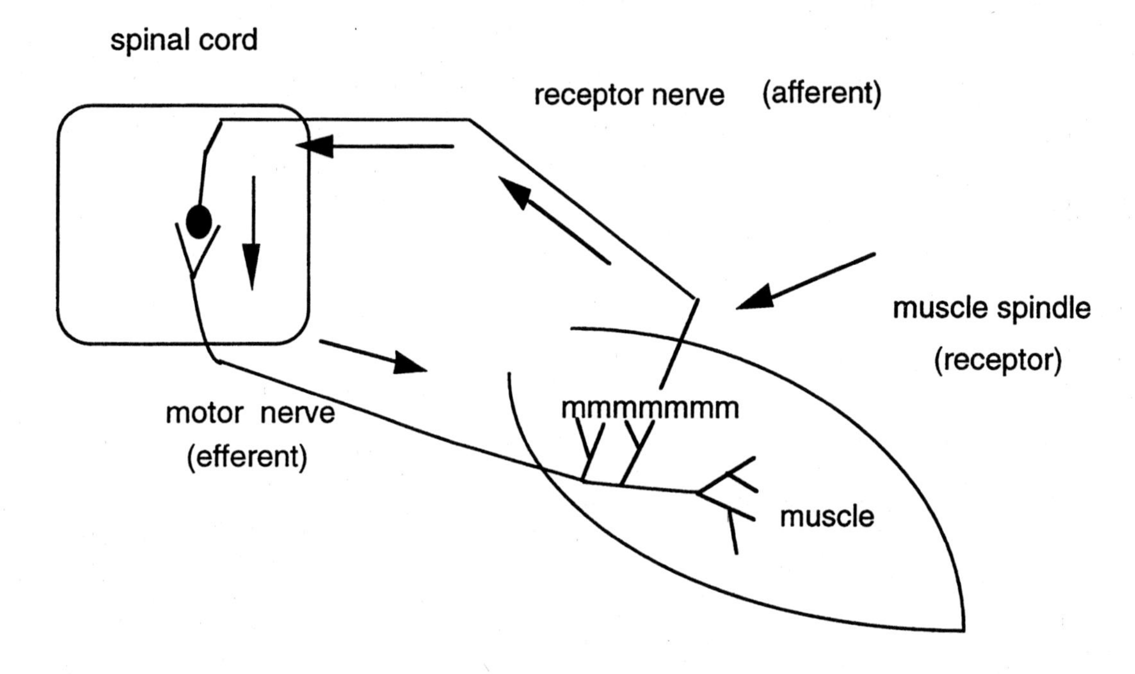

Tendons have receptors that detect muscle tension; namely, the **golgi tendon organs (GTO)** located in series with the fibers of the tendon. When the golgi tendon organ senses that the muscle is tense, it reflexively causes the muscle to relax. This is known as the **lengthening reaction**, which prevents the muscle from overcontracting and keeps the stress exerted on the bone to a minimum. The force created by the muscle during contraction causes its own inhibition or relaxation. It is possible for highly trained athletes to break their own bones because their muscles are so strong. If a muscle is "tight" or sore, it is soothing to massage the tendon. The GTO will cause the muscle to relax. Both the GTO and the muscle spindle respond to stretch but have opposite effects.

Table 1. Comparison of muscle and tendon receptors

Golgi Tendon Organ	Muscle Spindle
Fibers arranged in series with the tendon	Fibers arranged in parallel with the muscle fibers
Stretches when the muscle contracts	Stretches when the muscle is stretched
Inhibits motor neurons to the muscle so that the muscle relaxes	Stimulates motor neurons for the muscle to contract

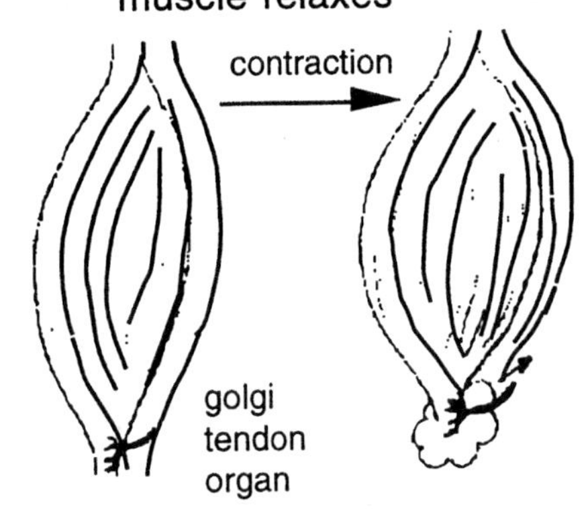

Golgi tendon organ response to muscle contraction

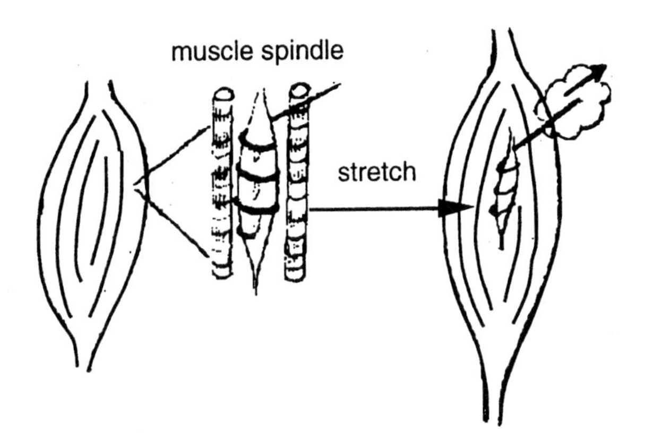

Muscle spindle response to stretching of muscle

Reflexes are often **reciprocally innervated** (Figure 3). Muscles with opposing actions are called *antagonists*. Through reciprocal innervation, the same interneurons (intermediate neurons located between the afferent and efferent neurons in the spinal cord) inhibit the action of one muscle while activating its antagonist muscle to provide the desired muscle action. For example, in arm flexion, the biceps (flexors) are activated and at the same time the triceps (extensors) are inhibited.

Reflex pathways that include one or more interneurons that are *polysynaptic* reflexes. Some interneurons form pathways that cross the spinal cord. These interneurons innervate extensor motor neurons (efferent neurons) on the opposite side of the body. When you walk, the extensors in your right leg are inhibited as your left leg swings forward. The arms are innervated the same way. This response is known as the **crossed extensor reflex.**

Figure 3. Reciprocal innervation of antagonist muscles

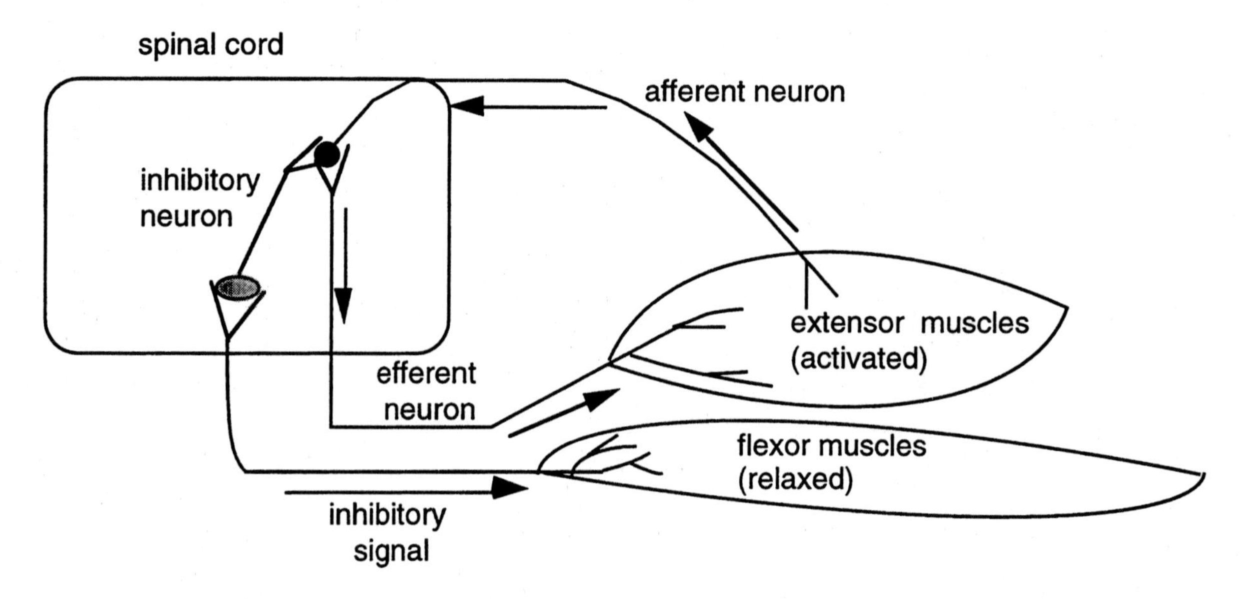

Differences exist between the reflex capabilities of endurance-trained and sprint-trained athletes. Generally, endurance-trained athletes have slower reflexes. Sprint-trained athletes have a slightly faster reflex time than nontrained individuals. One theory states that sprinters' muscle spindles are more sensitive to stretch. An equal tap on the tendon will cause less stimulation to a distance runner's muscle spindle than to a sprinter's. Researchers have found that ballet dancers exhibit less of an Achilles' tendon reflex than the average person because dancers must have fine, controlled movements; if their reflexes are highly active, their movements will be jerky.

Think of how swiftly you withdraw your hand after you have touched a hot pan. This is actually called the **withdrawal or flexor reflex**. It is activated most forcefully by feelings of pain. The withdrawal reflex is a polysynaptic reflex that uses reciprocal innervation circuits.

The speed of nerve conduction is increased with an increase in temperature. Nerve conduction is a series of chemical reactions, and all chemical reactions speed up with an increase in temperature. Similarly, cooler tissue temperature results in slower nerve conduction. There is also a decreased production of acetylcholine at low temperatures. Acetylcholine is the neurotransmitter released at neuromuscular junctions.

The brain is not part of the reflex circuit; however, the brain does modify the reflex. For example, a person who suffered brain damage during a car accident will have overly exaggerated reflexes because the brain can no longer contribute any inhibition or dampening of the reflex. The brain maintains a constant inhibitory effect on the reflexes. Often, when a doctor is checking a patient's reflexes, he or she will ask the patient to clasp both hands together and try to pull them apart. This is done to preoccupy the brain and therefore reduce the inhibition of the reflex. In much the same way, you are able to block out the ticking of the clock in class while listening to your teacher lecture.

A number of reflexes associated with the eyes can be readily demonstrated. One reflex involves the constriction of the pupil when a bright light is shined into the eyes. The energy of the light initiates the reflex by stimulating the parasympathetic nerves that control the pupillary sphincter muscles. In the absence of light, this reflex is inhibited and the pupils dilate. Another similar reflex is the **consensual light reflex** in which both pupils constrict. Even though the light is directed at only one eye, both pupils will constrict because commisural fibers in the optic pathway connect both eyes; therefore as one eye is stimulated to constrict, the other eye also receives the stimulus.

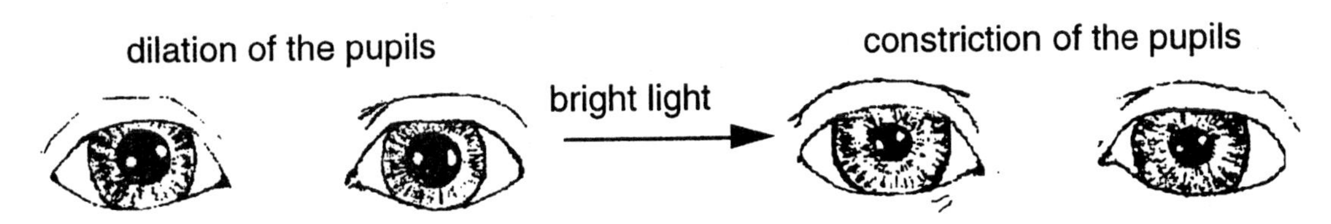

A second reflex involving the eyes is called **accommodation**. It is related to the focusing mechanism of the eyes. When the eyes focus on a distant object, the pupils dilate. If the eyes then shift to a nearby object, the pupils will constrict.

The taste buds send messages to the brain, which then enables us to experience taste. There is a reflex that triggers salivation when food is ingested. Have you ever had cravings for a certain food? Those cravings are not coincidental. Our bodies have adapted so well that they can sense our nutrient deficiencies and cause us to be "hungry" for foods that contain those nutrients. It has been shown that animals (including humans!) are fairly accurate in choosing food that fills their nutritional needs.

Nystagmus, the rhythmic oscillation of the eyes, is normally found in all humans. It occurs as the body tries to maintain a focus while the body is moving. In nystagmus, the semicircular canals within the ear are the receptors. After a person has been spun around in a chair, inertia causes the semicircular fluid to keep moving around even though the body has stopped moving. This can be demonstrated by swirling water around in a 2-L bottle. When you stop swirling the bottle, notice that the fluid continues to swirl around. This is the same principle that operates within the semicircular canals.

Each ear has three semicircular canals: horizontal, anterior, and posterior. All three are positioned at 90° to each other to represent the three-dimensional world we live in. The semicircular canals help keep our balance.

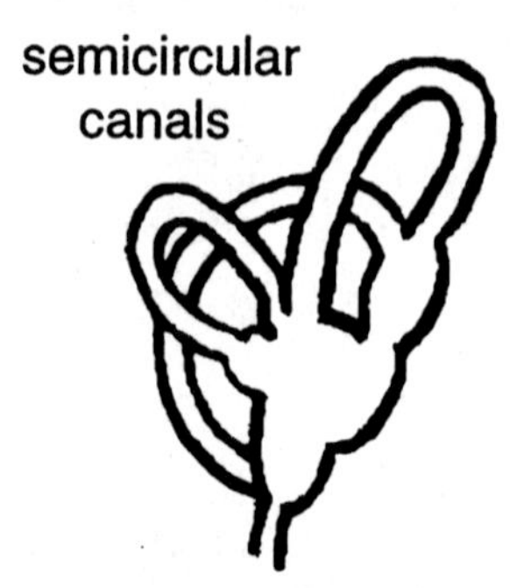

One of the experiments you will perform involves the nystagmus response. A student will spin around in a chair for 30 seconds. Another student will observe the oscillation of that student's eyes once the chair has stopped spinning. In the experiment, the canals sense that the body is still moving because they detect the fluid within the canals moving. If the canals sense the body is moving, the eyes must move. There is a fast and a slow eye component to nystagmus. The fast movement of the eyes is always in the direction opposite the movement of the fluid. Which way does the fluid in the semicircular canals flow? When a person is spun to the right, the fluid is moving to the right. Therefore the fast movement of the eyes is to the left. Spinning to the right is analogous to a sudden movement of the head to the left. In response to both actions, the fast movement of the eye is to the left. The body is simply trying to maintain a focus on a certain object. If the brain senses that the body is moving to the right, then the brain moves the eyes quickly to the left to go back to the point that it saw a short time ago. Think of riding a bus and seeing a friend on the sidewalk. As the bus is moving forward you have to look back to see your friend again.

Examples of nystagmus:

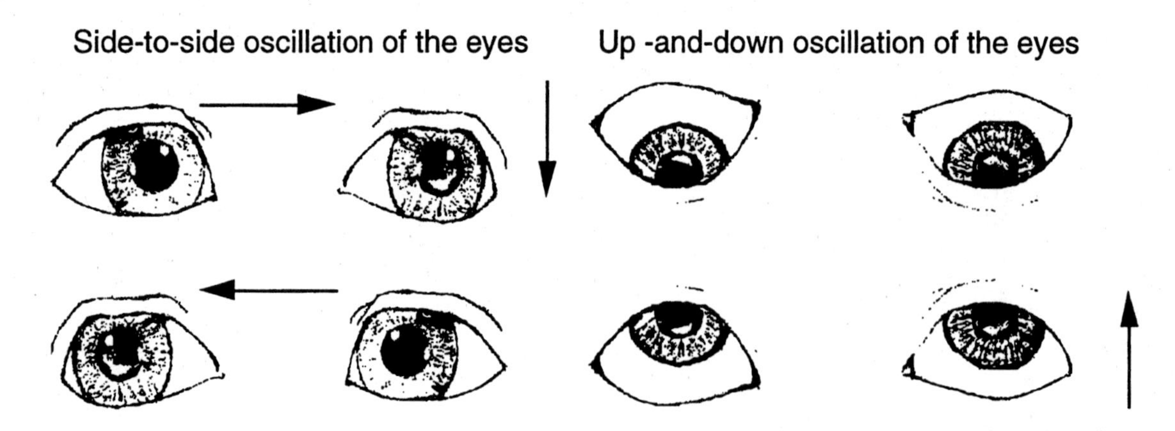

Part I: PATELLAR TENDON REFLEX

This exercise is designed to study the effects of temperature on the patellar tendon reflex as well as the inhibitory influence of the brain on the patellar tendon reflex.

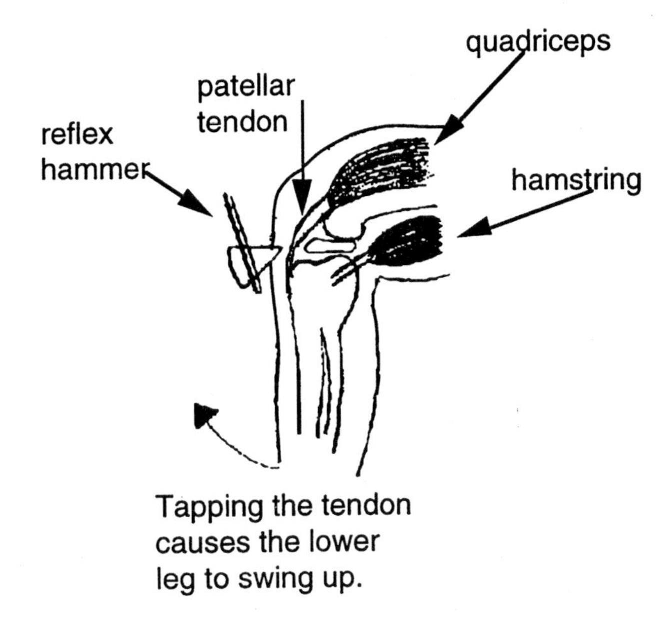

Tapping the tendon causes the lower leg to swing up.

PROCEDURE

1. Have the student sit on the edge of a table and let his or her legs hang freely.

2. Feel the tendon just below the knee; tap it with the reflex hammer or the side of your hand.

3. Record your observations (time of reflex, extent to which the leg kicks out). This is the control study; further observations will be compared with the control.

4. Wrap the student's thigh in a heating pad set on medium heat. After 10 minutes, tap the tendon with the hammer. Record your observations in Table 2.

5. Remove the heating pad and let the student's leg return to room temperature (about 10 minutes).

6. The student should clasp both hands together and try to pull his or her arms apart.

7. Repeat the procedure of tapping the patellar tendon; record your observations in Table 2. Be certain that the subject is trying to pull his or her arms apart at the time of the tendon tap.

8. Have the subject relax his or her arms. Now have the student concentrate on the reflex itself, thinking about his or her leg moving outward.

9. Repeat the tendon tap; record your observation in Table 2.

10. Have the student concentrate on the following numbers below and perform the addition in his or her head.

$$9 + 6 + 4 + 3 + 4 + 2 + 5 + 1 =$$

11. Repeat the tendon tap procedure while the subject is performing the addition; record your observations in Table 2.

12. Apply a cold pack to the subject's leg. Replace the ice as it melts.

13. Wait 5 minutes and then tap the tendon. Record your observations in Table 2.

14. The final exercise will compare reflexes and voluntary movements. Have the student kick out his or her leg when he or she hears a stimulus. The stimulus can be a whistle, snap of the fingers, clap, or tap on the table by another student. The student performing the voluntary movement should not be able to see the student who will provide the sound. Record the observations in Table 2.

Table 2. Patellar tendon reflex

ACTION	OBSERVATION
Initial tapping (control)	
Application of ice	
Application of heat	
Arms outstretched	
Concentrating on reflex	
Concentrating on problem	
Voluntary action in response to sound stimulus	

ANALYSIS OF PART I

1. Under what conditions did you observe a faster reflex? Why?

2. When did you observe a slower reflex? Why?

3. How did the voluntary action compare with the patellar tendon reflex?

Part II: DOORJAMB REFLEX

This exercise is designed to demonstrate the activity of the muscle spindle by having a student stand in the doorway and extend both arms to the side, pressing against the door frame. Maintaining a constant contraction in the arms does not affect the muscle spindle; however, when the student steps out of the doorway, gravity causes lengthening and therefore firing of the muscle spindle to cause muscle contraction.

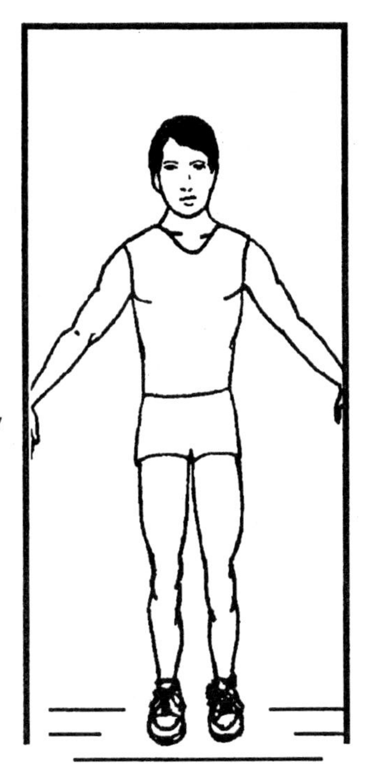

1. Have the subject stand in a doorway. With arms extended at his or her side, have the subject raise his or her arms until they touch the doorway.
2. Ask the subject to press hard on the door posts with both arms for 1 minute.
3. Have the subject walk forward with arms held in the same position. Record what happens.

   OBSERVATIONS:

ANALYSIS OF PART II

1. What happened to the subject's arms when he or she walked away from the doorway?

2. Why does this occur?

Part III: OCULAR REFLEXES

This exercise is designed to demonstrate the reflexes that occur in the eye upon sudden changes in light exposure and viewing distance.

A. Pupillary Constriction

PROCEDURE

1. Choose a partner. Have the partner close his or her eyes while sitting in front of a 60-W light bulb.

2. When your partner opens his or her eyes, watch for any changes in the eyes. Record your observations in Table 3.

B. Pupillary Accommodation

PROCEDURE

1. Choose a partner and have them observe the position of your eyes.

2. Focus on a distant object over 20 ft away. Your partner should record observations in Table 3.

3. With no change in lighting, place your finger about 10 in. from your face and look at it. Your partner should record observations in Table 3.

4. Bring your finger in closer and closer until the right eye is no longer able to focus on the finger. Hold your finger still at this point and measure this distance between the right eye and the finger. This is your *near point.* Repeat the activity for the left eye and record the data below.

   Right eye near point: ____ cm   Left eye near point: ____ cm

C. Consensual Light Reflex

PROCEDURE

1. Choose a partner. Place your hand between the subject's eyes, in line with the subject's nose. Shine a light in one eye only. Record observations of both eyes in Table 3.

Table 3. Ocular reflexes

ACTION	OBSERVATION
Response to bright light	
Response to distant object	
Response to nearby object	
Response of both pupils to light shone in one eye	

ANALYSIS OF PART III

1. Why did the changes in the pupil occur upon exposure to light?

2. What were the differences in the position of the eyeball and in pupil size when the eye was focusing on a far object? A near object?

Part IV: THE CROSSED EXTENSOR REFLEX

This exercise is designed to demonstrate how "natural" our reflexes are.

1. Walk about 10 ft, noting what it feels like.

2. Walk an additional 10 ft except this time moving your right arm forward with the right leg and swinging the left arm forward with the left leg. Do these actions feel unnatural?

Part V: TASTE REFLEX

This exercise is designed to demonstrate the taste reflex.

1. Prepare a drink from dry powder concentrate. (EX: lemonade or iced tea)
2. Drink a glass of the beverage at normal concentration.
3. Drink a glass of the beverage at half the concentration.
4. Drink the beverage at three times the normal concentration. Record your observations and comments.

OBSERVATIONS:

Part VI: NYSTAGMUS REFLEX

This exercise is designed to demonstrate the oscillation of the eyes that occurs when an individual is spun around.

PROCEDURE

1. The student should sit in the center of the room on a swivel chair. (Other students should be close by in case the subject falls out of the chair.)
2. Have the student bend his or her head forward slightly (30°).
3. Spin the chair rapidly in a clockwise direction for 20–30 seconds.
4. Stop the rotation of the chair suddenly and observe the student's eyes. Record your observations.

   OBSERVATIONS:

5. Repeat the same experiment with the student's head tilted at an angle of 90° forward throughout the spinning. When the chair is stopped, the subject should sit upright; observe his or her eyes.

   OBSERVATIONS:

ANALYSIS OF PART VI

1. Why is it important to tilt the head forward slightly?

2. What was the difference observed when the head was tilted 90°? Why?

LABORATORY EXERCISE 8

BIOMECHANICAL AND LENGTH-TENSION RELATIONSHIPS OF SKELETAL MUSCLE

BACKGROUND AND KEY TERMS

Skeletal muscle is very organized tissue. The whole muscle is composed of individual muscle fibers, which are composed of smaller units called *myofibrils.* The components responsible for muscle contraction are the *actin* and *myosin* filaments that make up the myofibrils. Each myofibril is made up of about 1500 myosin filaments and 3000 actin filaments. The organization of actin and myosin filaments forms a banded pattern that can be seen under a microscope. The darker bands represent the area of the myofibril that contains actin and myosin side by side. Light bands represent the area of the myofibril that contains only actin filaments. Actin and myosin are polymerized proteins; their interaction produces muscle contraction.

Figure 1. Myofibril section

Section of myofibril that shows the thick myosin filaments and the thin actin filaments

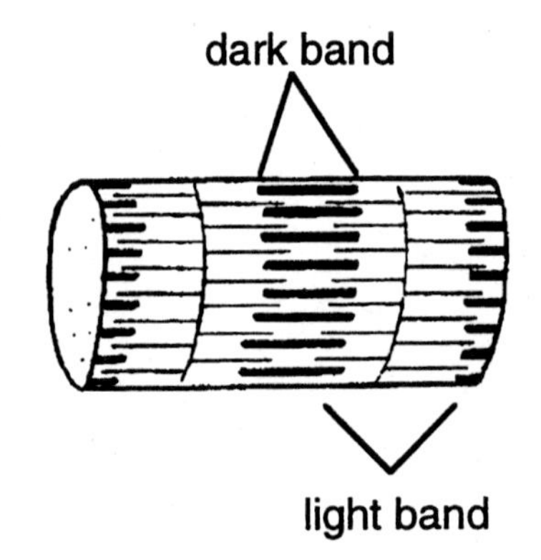

There are two main components to muscle strength. The first is the length-tension relationship that is based on the interaction of the microscopic actin and myosin fibers. The biomechanics of the musculoskeletal system is the second component. The length-tension curve (Figure 3) demonstrates that each muscle fiber has an optimal length for generating maximal force. However, this length does not correspond to the most advantageous position according to the biomechanics of our muscles, bones, and joints. The following exercise will examine how these two factors affect our movements and strength.

Muscle contraction is explained by the *sliding filament theory*. Cross-bridges originating on the myosin filament attach to the actin. The cross-bridges pull the actin filaments across the myosin in a step called the *power stroke*. The actin filaments slide over myosin, thereby shortening the muscle fiber and producing muscle contraction (Figure 2). During the contraction, myosin filaments remain stationary.

Figure 2. The sliding filament theory of muscle contraction in which actin filaments slide over myosin filaments

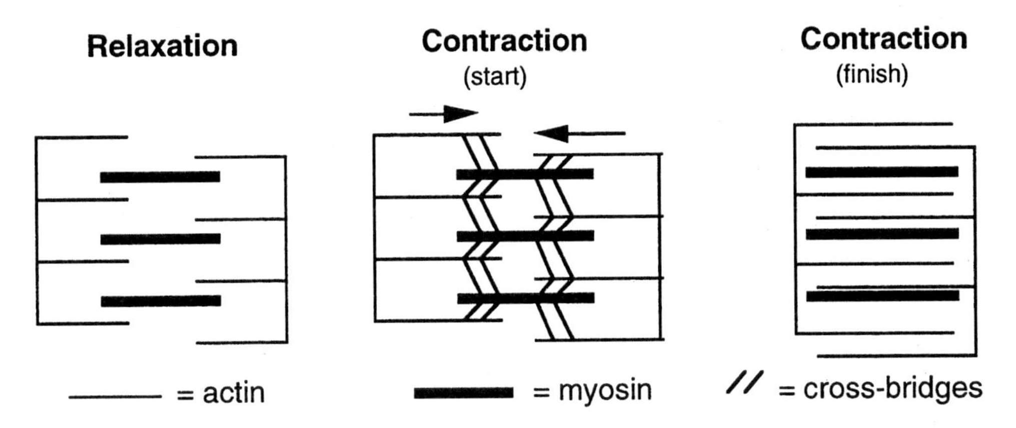

Maximum force is generated when the maximum number of cross-bridges between actin and myosin are formed. The cross-bridge formation corresponds with the length-tension curve of a single muscle fiber in Figure 3, where L_{max} represents the length at which the greatest tension is produced. This graph represents the muscle fiber only in vitro (out of the body). The muscle within the body is limited by surrounding bone and connective tissue from stretching beyond L_{max}. Therefore, only the ascending part of the length-tension curve describes the force-generating characteristics in the muscle fiber in vivo (within the body).

Figure 3. Length-tension curve of a muscle fiber. L_{max} represents the optimal length of the fiber for producing maximal tension. The greatest number of cross-bridges are formed between actin and myosin filaments at L_{max}.

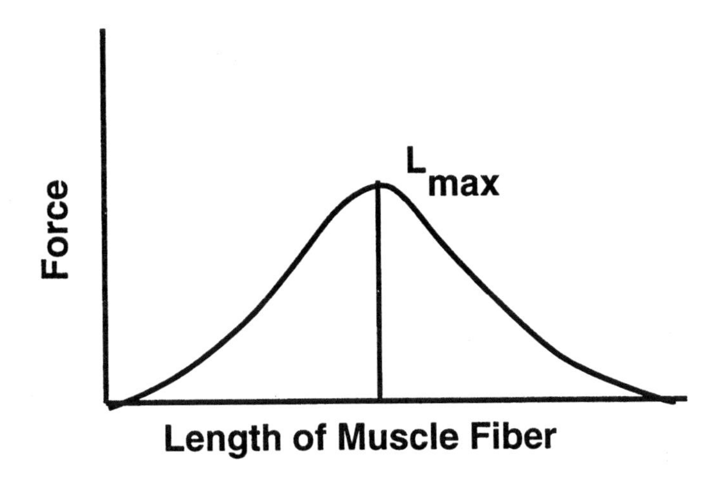

Weight lifters utilize the length-tension relationship. Their muscles increase in size due to the existence of greater numbers of myofibrils. Since myofibrils consist of actin and myosin, the muscles of the weight lifter have more contractile proteins to interact with each other. More cross-bridges form during muscle contraction and therefore more force is generated in the muscle.

Active insufficiency occurs when a muscle shortens to a length at which it can no longer create tension. A muscle fiber can contract to only about half of its length. Active insufficiency occurs only in muscles that cross more than one joint. The flexor muscles of the ventral (underside) portion of the forearm are an example of muscles that demonstrate active insufficiency. They cross the wrist joint and also the joints of the fingers. When the flexor muscles contract to flex the wrist, the flexor muscle shortens partially. It is difficult to make a fist while your wrist is flexed because the flexor muscles required to form a fist are already shortened to flex the wrist; therefore only minimal shortening of the flexors can occur to make the fist. However, if the wrist is extended, you can make a powerful fist. The flexor muscles have two main functions: flexing the wrist and flexing the fingers (making a fist). Because the contraction has limited range, both actions cannot be performed simultaneously with maximum force.

Passive insufficiency occurs when the muscle fiber can no longer lengthen. As in active insufficiency, this limitation also occurs in muscles that cross more than one joint. The difference is that the insufficient muscle is not the muscle that is contracting. In the following example, the main contracting muscle is the anterior tibialis (in the front of the lower leg), and the calf muscle is the insufficient muscle. The calf muscle (gastrocnemius) crosses the knee joint and the ankle joint. When the lower leg is extended, the foot can be pulled upward (dorsiflexed) only a small amount. The calf muscle is stretched fully when the leg is extended. It acts as a limiting factor in preventing the foot from pulling upward because the calf muscle can stretch no further. Therefore, when the leg is bent, the calf muscle is slack and the foot can dorsiflex a greater amount.

Have you ever noticed that as you screw a screw in with a screwdriver, the harder you struggle, the more often the screwdriver pops out of the groove on the screw? This occurs because the biceps muscle serves two functions: it primarily flexes the arm and secondarily supinates the arm (turns the arm from palm side down to palm side facing up). When turning the screwdriver with force, you are mainly using the supinator function of the muscle. But because you are strongly contracting your biceps, its primary function, elbow flexion overrides the secondary function, and the screwdriver pops out of the screw. This is common in muscles with secondary functions. When using a muscle forcefully in its secondary motion, the primary motion will often take over, just as in the screwdriver example.

Synergy is a combined action to achieve a result not possible when each of the components acts alone. Muscles can act as synergists. Action is possible when two or more muscles act together; if one of the muscles is damaged, the intended action cannot be performed. The muscles have actions that they share and also actions that are opposite (antagonistic) one to another. The antagonistic actions "cancel" each other out–neither action is observed. The action of extending your wrist upward is the result of synergistic muscles. The muscles involved are the extensor carpi radialis and the extensor carpi ulnaris. The extensor carpi radialis extends the wrist and radially deviates it. Radial deviation is movement of the wrist toward the this muscle causes the left wrist to rotate to the right. Similarly, the extensor carpi ulnaris muscle extends the wrist and ulnar deviates the wrist (turns it outward). When working together, these two muscles extend the wrist straight up.

Biomechanics is the study of the human body in terms of forces, levers, friction, and the like, using principles of Newtonian physics to analyze human motion. A simple way to explain biomechanics as it relates to the biceps muscle is with the following equation:

$$F \times FA = R \times RA$$

where F = force needed
FA = perpendicular distance from the joint to the muscle insertion
R = resistance (the amount of weight held)
RA = the perpendicular distance from the weight (R) to the joint.

The equation is less complicated than it looks. See Figure 4 for a visual representation of the equation.

Figure 4. Components of the equation $F \times FA = R \times RA$

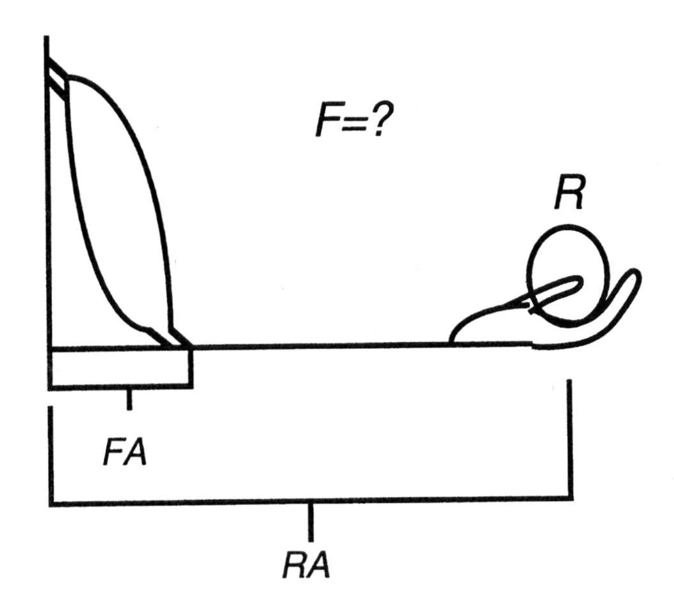

For example, if the length of the subject's forearm is 36 cm, and the biceps muscle inserts 5 cm from the elbow, we know that RA = 36 cm and FA = 5 cm.

The 110-kg weight that the subject is holding equals the resistance: R = 110 kg. The force *(F)* required to lift the weight is the unknown. If we insert the values into the equation, we get

$$F \times 5 \text{ cm} = 110 \text{ kg} \times 36 \text{ cm} \quad \text{or} \quad F = \frac{110 \text{ kg} \times 36 \text{ cm}}{5 \text{ cm}}$$

Solving for force, we find that 792 kg of force is necessary to lift the 110-kg weight.

When the subject's forearm is raised at a 45° angle (Figure 5), both the *FA* and the *RA* decrease while *R* remains the same (the subject lifts the 110-kg weight again). The insertion remains exactly the same distance from the joint, but the force arm changes. The force arm changes because the force arm is determined by the *perpendicular* distance from the joint to the insertion. The length of the resistance arm changes also, because the resistance arm is the perpendicular distance from the joint to the weight. The new values are found to be *FA* = 4.3 cm and the *RA* = 26.7 cm. After inserting these new values into the preceding equation, we find that the new force required is about 683 kg of force. Notice that it takes considerably less force to lift the *same* weight when the arm is at a different angle. At the new angle, the arm has a mechanical advantage. Biomechanics is used in Nautilus machines. These machines provide added resistance as the athlete goes through the midrange so that maximum exertion can be achieved throughout the *entire* range. In other words, the machine is set to provide more resistance at the angles at which a person is strongest.

Figure 5. Changes in the biomechanical components when the arm is raised to 45°. *FA* is the perpendicular distance from the joint to the insertion, and *RA* is the perpendicular distance from the joint to the weight.

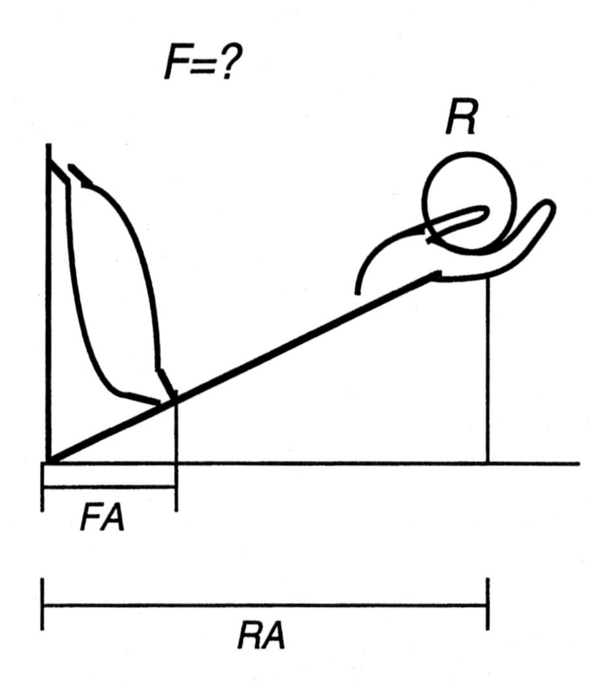

It is easy to see from the force equation that it is an advantage to have a muscle insertion farther from its associated joint. Males typically have muscle insertions that are farther away from the point of rotation; this is one of the many reasons why men are typically stronger than women.

Biomechanics looks at the engineering of the human body. A two-joint muscle such as the biceps muscle has orientations that are advantageous to the strength of the muscle. Flex your elbow and pronate your hand (rotate thumb down); now supinate your hand (rotate thumb out) and observe. The biceps is considerably larger when the hand is supinated because the biceps muscle is wrapped around the radius during pronation; therefore the biceps is at a mechanical disadvantage during pronation.

Part I: MEASUREMENT OF THE STRENGTH OF THE BICEPS MUSCLE AT VARIOUS ANGLES

This exercise is designed to illustrate the effect of varying the joint angle on the maximum weight that can be lifted at that angle.

PROCEDURE

1. With the aid of a protractor, mark a 60-cm wide by 30-cm high poster board with angles of 0°, 20°, 45°, 60°, 90°, and 120°. Mark the 90° angle exactly in the middle (30 cm from the edge), and mark the remaining angles starting with 20° from the left side of the poster board. Label the respective angles as shown in the figure.

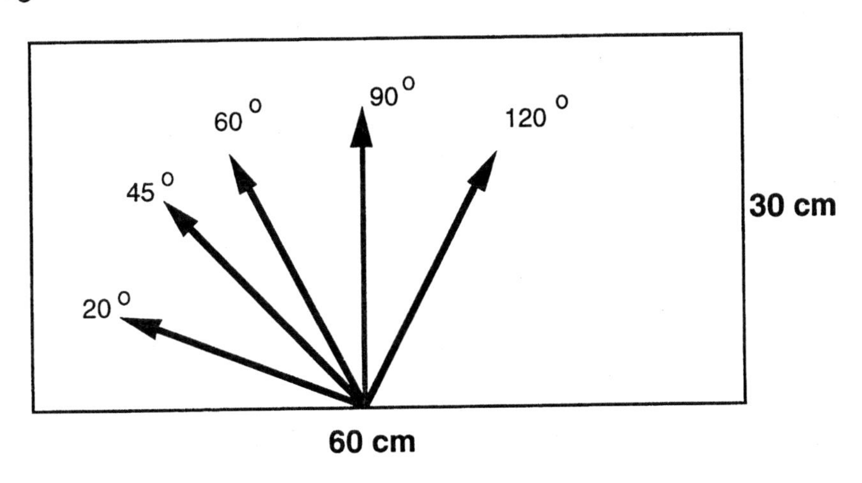

2. Have the subject place his or her right arm (shoulder to elbow) flat on a desk. Now line up the forearm (elbow to hand) with the poster board at an angle of 20°.

3. Estimate the maximum weight that the subject can hold at the specified angle. Place a dumbbell of a weight smaller than the estimated maximum in the subject's right hand. The subject should maintain his or her arm at the same starting angle even as weights are being added.

4. Add additional weight to the dumbbell (weights are plates that are added to the dumbbell) in the smallest increments possible until the subject can no longer hold the weight. Record the maximum weight that the subject can hold at the 20° in Table 1. Repeat the procedure for each angle labeled on the posterboard.

5. Repeat steps 2–4 with the left arm at each labeled angle. Record the maximum weight held at each angle. Plot the maximum weight at each angle for the right arm and for the left arm in Graph 1. Use a solid line for the right arm and a dashed line to represent the left arm.

ANALYSIS OF PART I

1. At which angle does the biceps muscle exhibit its greatest strength?

2. At which angle is the body at a mechanical advantage? Why?

3. At which angle is the body at a length-tension advantage? Why?

Table 1. Strength, force arm, resistance arm, resistance, and force values for various angles in the range of motion

Angle (°)	Max. wt. held w/right arm	Max. wt. held w/left arm	Force Arm (*FA*)	Resist-ance Arm (*RA*)	Force (*F*) needed to lift *R*	Max. wt. lifted 0° (*R*)	Force (*F*) required to lift weight
20							
45							
60							
90							
120							

GRAPH 1. Maximum Weight Lifted at Various Angles

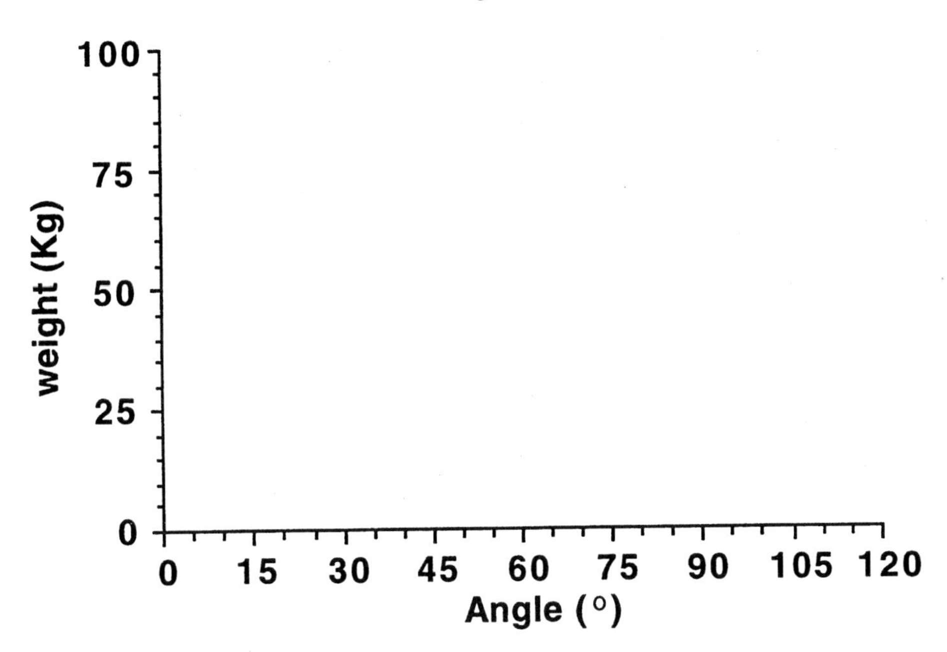

Part II: CALCULATION OF THE FORCE REQUIRED TO LIFT A MAXIMUM WEIGHT AT VARIOUS ANGLES

PROCEDURE

1. Measure the length of each subject's right forearm (*RA*) in centimeters. This is the length from the medial epicondyle, a bony prominence that can be felt on the inside of the arm at the elbow, to the middle of the palm. Place your thumb and forefinger on the bones on each side of your elbow; as you flex your elbow you should feel the bones moving. If your body were split into right and left halves, the midline would separate the right side from the left side. The bone on the medial side (toward the midline of your body) is the medial epicondyle. Enter the distance from the medial epicondyle to the middle of the palm as the resistance arm at the angle 0° in Table 1.

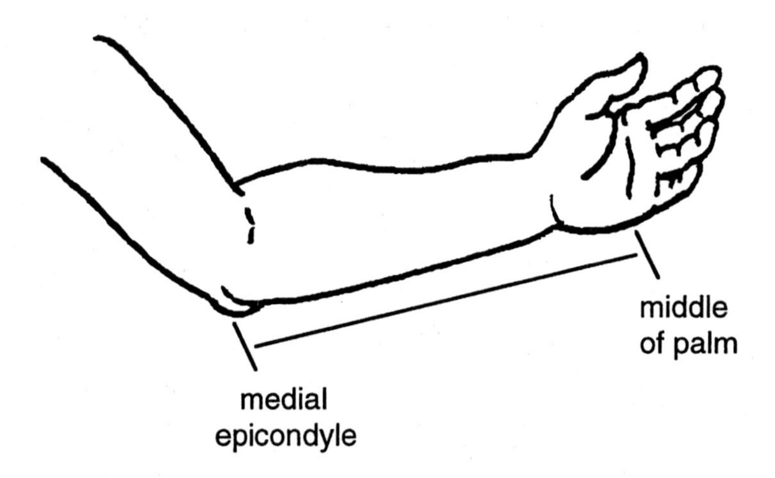

2. On Graph 2, plot the right forearm length on the *x*-axis. The *y*-axis represents the right arm. (Remember that the arm is the region from the shoulder to the elbow, and the forearm is from the elbow to the wrist).

3. It is estimated that the biceps insertion is around 2.54 cm from the elbow in females and about 5 cm from the elbow in males. Place a mark on Graph 2 to represent the insertion point.

4. Use a protractor to mark the angles 0°, 20°, 45°, 60°, and 90° on the graph. Line up the protractor with the *y*-axis so that 0° is at the *x*-axis and 90° is at the *y*-axis. Use a ruler and draw lines the length of the right forearm from the *x*-*y* intercept through each angle. All five lines should be the same length. See Figure 6.

Figure 6. Example of how Graph 2 should be drawn

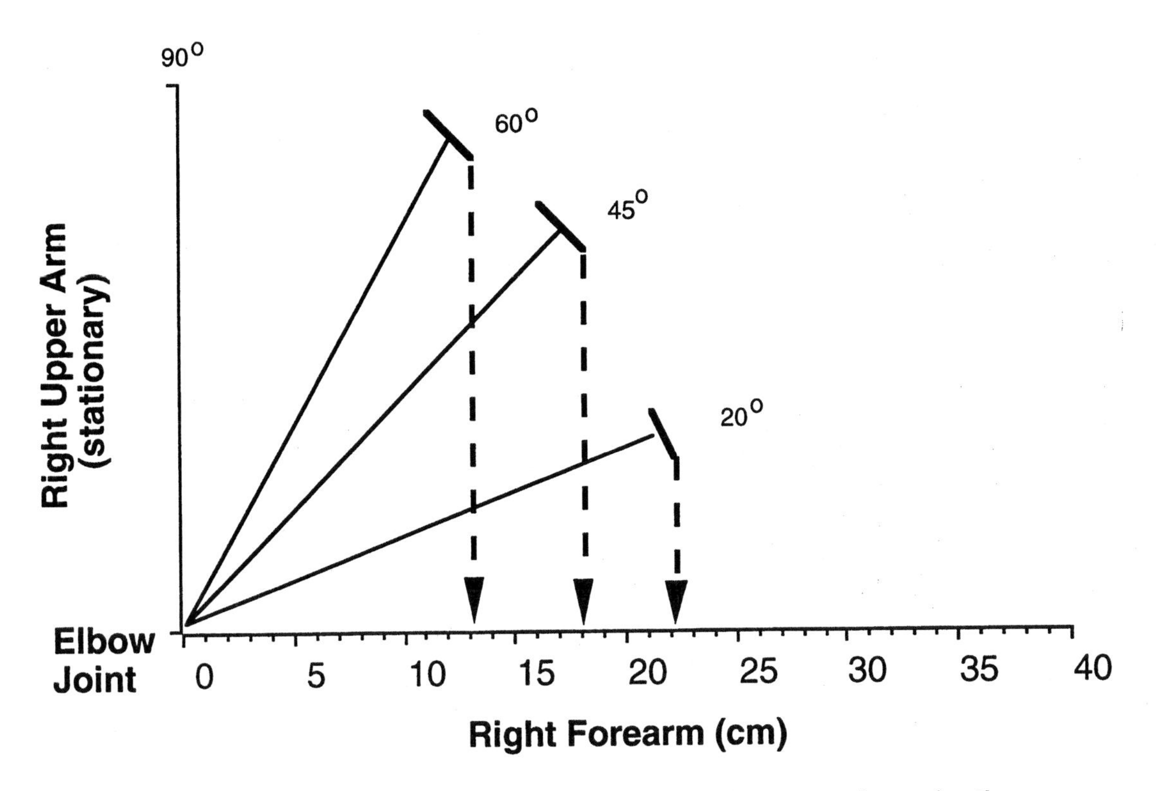

5. Use a ruler to mark the muscle insertion on the line at each angle; the distance from the joint to the insertion remains the same through all the angles. For example, if the subject is a female, mark the insertion 2.54 cm from the joint (*x-y* intercept =0) on the lines for each of the angles.

6. With the aid of a protractor, draw perpendicular lines from the muscle insertion to the *x*-axis. The line from the insertion point should intersect the *x*-axis at a right angle. The distance on the *x*-axis from the origin to the perpendicular line represents the force arm (*FA*). Record this measurement for each angle in Table 1.

7. Using the protractor, draw perpendicular lines from the endpoints of the five lines drawn through each angle. The distance on the *x*-axis from the origin to the perpendicular line represents the resistance arm (*RA*). Record these measurements in Table 1.

8. Compute the force (*F*) the subject needed to lift the weight at each angle. For the resistance (*R*), use the values entered in column 2. Enter the force needed at each angle in column 6.

Graph 2 . Change in reistance for right forearm at 0, 20, 45, 60, and 90°

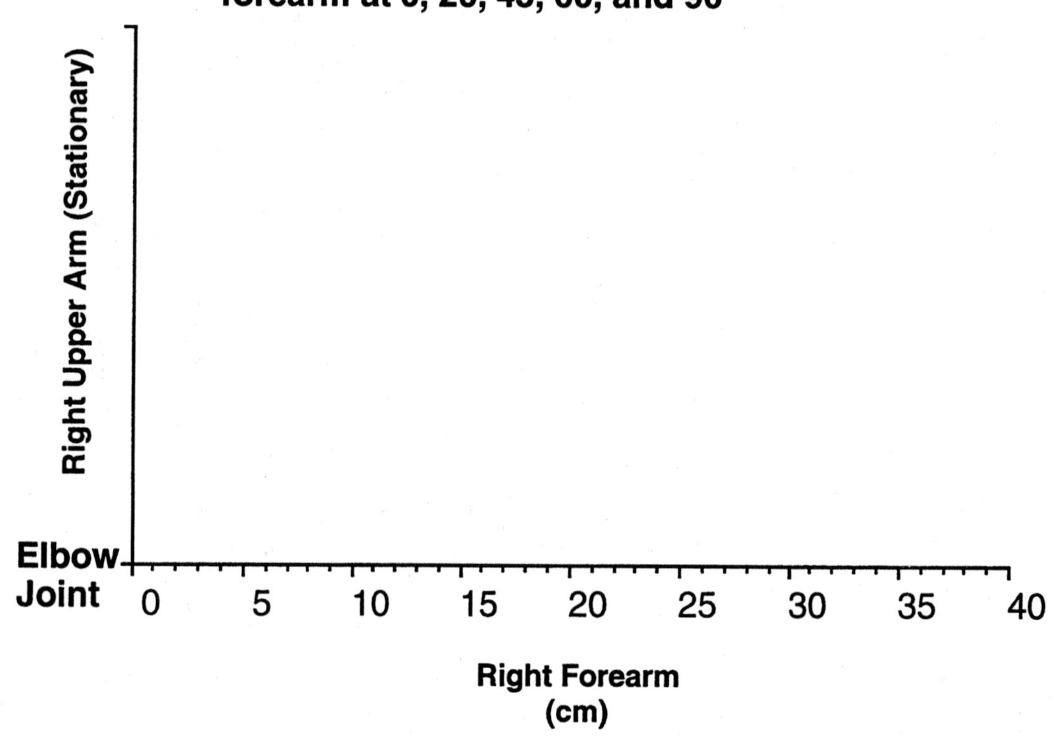

Part III: CALCULATION OF THE FORCE REQUIRED TO LIFT A GIVEN WEIGHT AT VARIOUS ANGLES

PROCEDURE

1. Use the values obtained from Part II. However, use the maximum weight lifted at 20° as the resistance at each angle.

2. Compute the force required to lift the weight (maximum weight lifted at 20°) through the range of motion and enter the force required in Table 1, column 8.

Part IV: ACTIVE INSUFFICIENCY

PROCEDURE

1. Have the subject extend his or her wrist and make a powerful fist. Record your observations.

2. Have the subject flex his or her wrist and make a powerful fist. Record your observations.

   OBSERVATIONS:

Part V: PASSIVE INSUFFICIENCY

PROCEDURE

1. Have the subject extend his or her leg. Hold a string from the side of the knee to the heel. Pull the string tight.

2. Have the subject bend his or her leg a small amount while the string is still in place. Notice what happens to the string. The string represents the events occurring in the calf muscle.

3. With leg extended, have the subject dorsiflex (pull toes back) the foot.

4. Have the subject bend the leg and dorsiflex his or her foot as far as possible. Record observations on the subject's range of motion in both positions.

   OBSERVATIONS:

LABORATORY EXERCISE 9

RENAL PHYSIOLOGY AND PRACTICAL APPLICATIONS OF STARLING'S LAW OF THE CAPILLARY

BACKGROUND AND KEY TERMS

The kidneys are two bean-shaped organs located in the back of the abdomen along each side of the vertebral column (Figure 1). The kidneys regulate water balance within our bodies. This is an important task, considering that 60% of the human body is made up of water. The kidneys also balance the concentration of electrolytes and ions within the body's water and control the excretion of metabolites. Some hormones are regulated by the kidneys.

Figure 1. Anatomy of the kidney

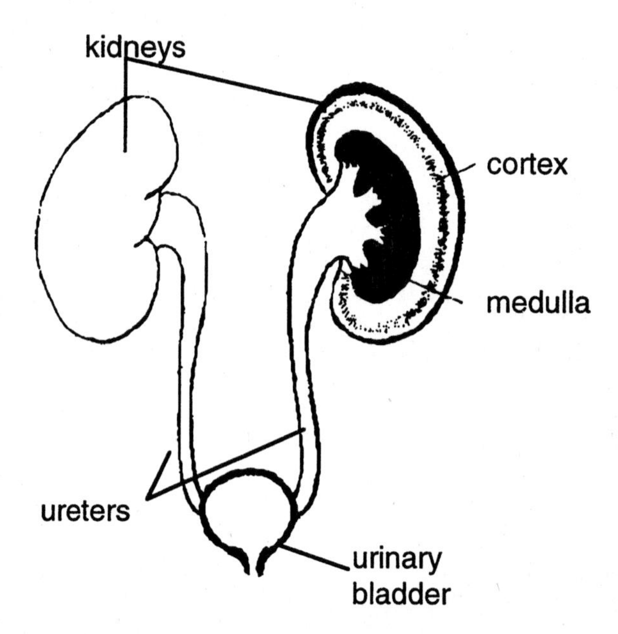

The primary functional unit of the kidney is the nephron, a microscopic filtering unit that removes unwanted substances from the blood. Blood enters the nephron through the **afferent arteriole** (Figure 2). It then passes through a network of capillaries called the **glomerulus**, where it is filtered. The fluid leaves the glomerular capillaries because the pressure inside the capillaries is high. The filtered fluid passes through a surrounding structure called **Bowman's capsule**. It then travels through a series of tubes that eventually lead to large collecting ducts. The filtered blood leaves the capillaries through the **efferent arteriole**.

Figure 2. The nephron is the functional unit of the kidney. The blood entering the kidney is filtered here.

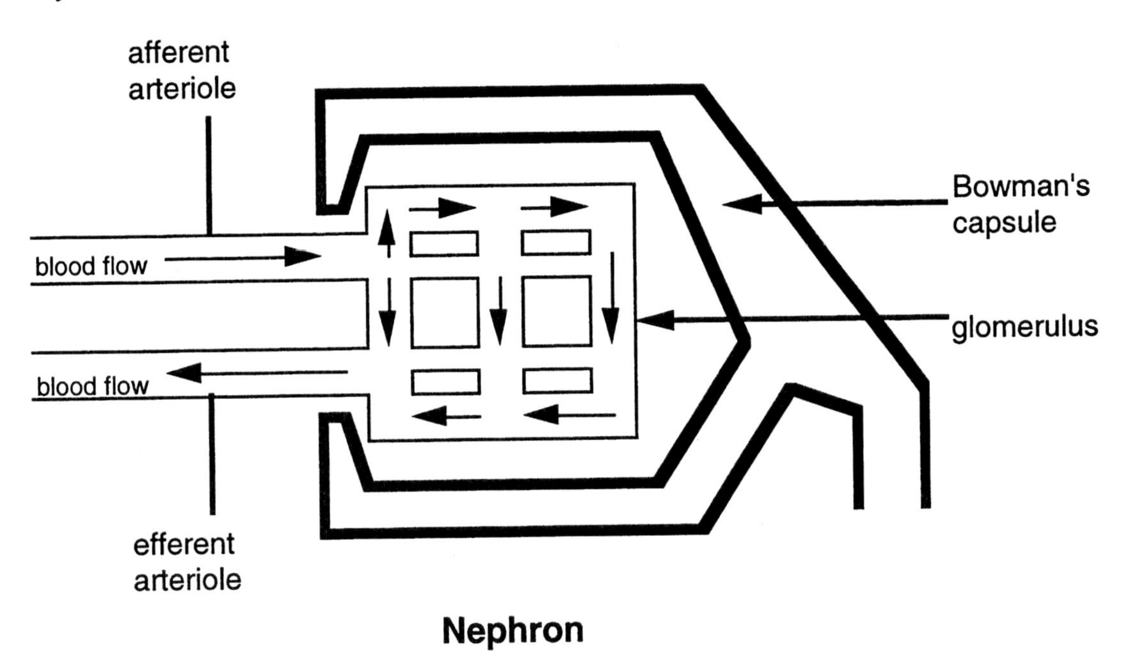

Nephron

Osmosis is the movement of water across a semipermeable membrane from a compartment with low solute concentration into a compartment with high solute concentration. Our kidneys function by this principle. The semipermeable membrane is the blood vessel wall, while the compartments are the interstitium (space between cells) and the blood vessels (capillaries). Water flows into the compartment with high solute concentration in an attempt to dilute it and equalize concentrations in both compartments, a state called **equilibrium**.

Starling's equation is used to determine the net movement of fluid and solutes across the capillary membrane. In other words, this equation determines the flow. Starling's equation is as follows:

$$\text{Flow} = K_f \; [(P_c - P_i) - (\Pi_c - \Pi_i)]$$

+ answer = flow out of the capillary

- answer = flow into the capillary

where P_c = capillary hydrostatic pressure

P_i = interstitial hydrostatic pressure

Π_c = capillary oncotic pressure

Π_i = interstitial oncotic pressure

K_f = filtration coefficient, or membrane permeability X membrane surface area

An understanding of the preceding equation requires an understanding of two different pressures: hydrostatic pressure and oncotic pressure. The first is **capillary hydrostatic pressure.** Think of this as similar to the pressure that forces water to squirt out of the side of a garden hose that has a hole in it. This pressure is created by the fluid pushing against the sides of the vessel; this pressure forces fluid out of the capillary. The second type of pressure is **capillary oncotic pressure**, which ultimately pulls water into the vessel. Capillary oncotic pressure is due to the presence of nondiffusible proteins in the blood vessel. The proteins cannot pass through the capillary membrane; therefore they produce an increased solute concentration inside the capillary. Water moves into the vessel by osmosis due to the high solute concentration inside the vessel. In addition to capillary pressures, there exist hydrostatic and oncotic pressures in the interstitium. Capillary hydrostatic pressure opposes capillary oncotic pressure. There is a battle between the forces pushing water out of the capillary and the forces pulling water into the capillary. The Starling equation incorporates all these forces and predicts which forces will win.

Figure 3. Pressures along the glomerular capillary responsible for filtration

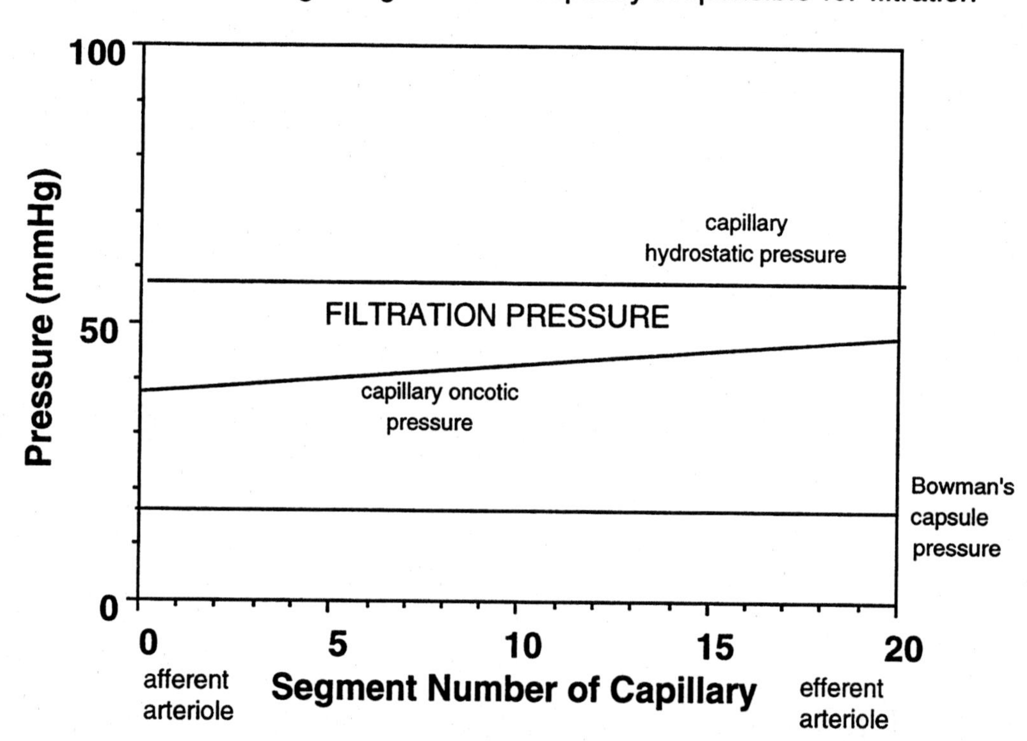

Imagine that one of the coiled capillaries (glomerular capillary) could be spread out into a continuous straight vessel and divided into 20 segments (Figure 3). The pressure in each individual segment was determined and plotted on the graph. The capillary hydrostatic pressure remains almost the same throughout all segments of the vessel because there is little resistance to decrease the blood flow. The low resistance is due to the fact that the glomerular capillaries are arranged in parallel. As the capillary oncotic pressure increases toward the efferent arteriolar end it reduces the filtration pressure; filtration pressure is responsible for forcing plasma out of the vessel. Oncotic pressure increases (increase in concentration of proteins within the capillary) and therefore pulls fluid into the capillary to dilute the protein concentration. So even though the force pushing the plasma out of the vessel (capillary hydrostatic pressure) remains essentially the same, the force pulling water into the capillary (oncotic pressure) increases. The net result is a lower filtration pressure and reduced flow out of the capillary.

The hydrostatic capillary pressure wins out in the afferent end of the glomerular capillary. The body naturally forces as much fluid as possible out of the capillary for the kidneys to filter. As water is being forced out of the capillary, electrolytes, amino acids, glucose, and other nutrients follow the water into the interstitium (between the capillary and the cells of the kidney tubule). If these nutrients remain in the interstitium they will be filtered and excreted by the kidneys, resulting in the loss of essential nutrients. Only a small portion of the plasma, filtered by the kidney, is excreted as urine. The capillary oncotic pressure is higher in the efferent end of the arteriole and pulls water and nutrients back into the capillary so they can be utilized by the tissues.

Figure 4. Plasma flow from the afferent to efferent end of the capillary

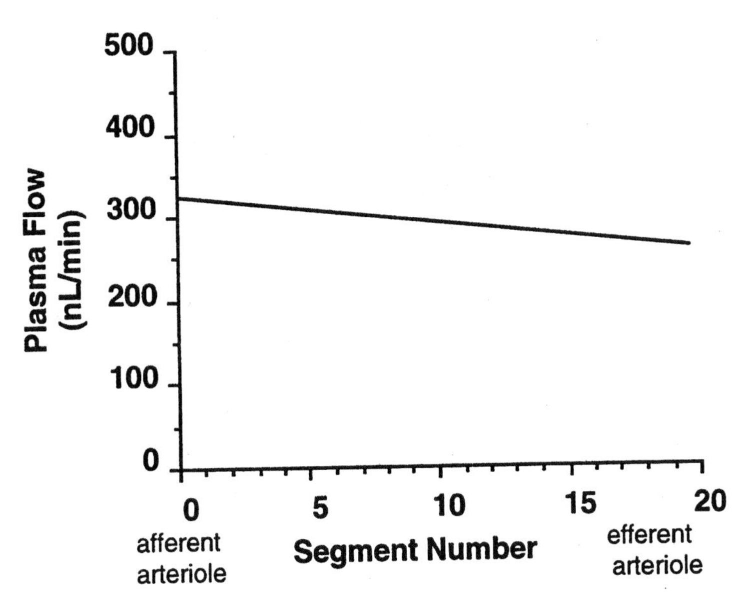

As can be seen from Figure 4, plasma flow decreases as the blood moves toward the efferent arteriole. Flow is dependent on the pressure inside the tube and the resistance (flow = pressure / resistance). As fluid is filtered out of the capillaries, the capillary pressure decreases; therefore flow decreases. This is because as the fluids move out, the solutes remaining in the vessels grow more and more concentrated. This lessens the gradient that forces the plasma out of the arteriole. Because the filtration rate is directly dependent on the plasma flow, there is a similar drop along the capillary (Figure 5).

Figure 5. Filtration rate from the afferent to the efferent end of the capillary

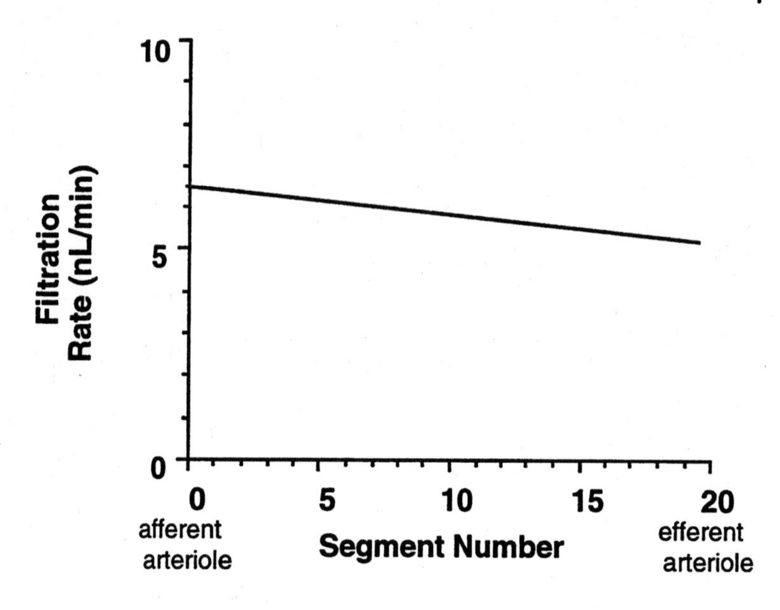

Part I: RENAL PHYSIOLOGY

This exercise is designed to examine the effects of exercise, low blood volume, dehydration, and high salt intake on the parameters of the Starling equation.

PROCEDURE

Table 1 illustrates what happens to capillary filtration (GFR) when changes occur in the capillaries. Using the table, predict what will happen to filtration (GFR) under the following conditions.

Table 1. Effect of dilating and constricting the efferent or afferent arteriole end of the glomerular capillary.

	Renal Blood Flow (RBF)	Glomerular Filtration Rate (GFR)
Constrict afferent arteriole	↓	↓
Dilate afferent arteriole	↑	↑
Constrict efferent arteriole	↓	↑
Dilate efferent arteriole	↑	↓

1. The Effects of Exercise

When Hector plays tennis, his kidneys function differently. During exercise, the blood is shunted away from the splanchnic organs such as the spleen or kidney to provide the active muscles with more blood. To do this, the body constricts the afferent arteriole while the glomerular filtration rate (GFR) remains the same.

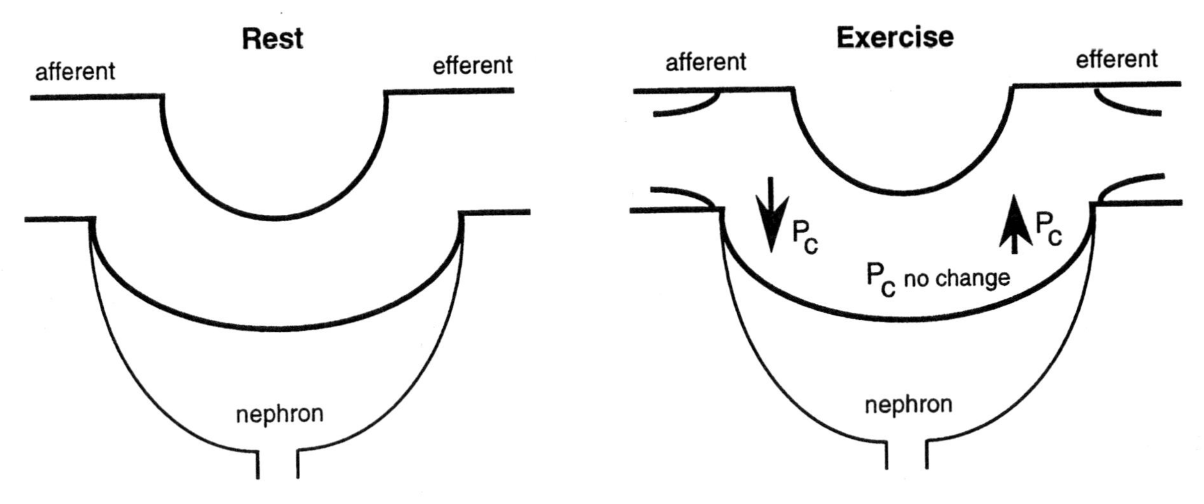

a. How does the body maintain the same GFR during exercise?

b. What would result if the body did not shunt the blood away from the kidneys during exercise?

2. The Effects of Low Blood Volume

Steve was walking to his female friend's house when he was attacked and bitten by two pit bull dogs that had escaped. He managed to survive, but he lost much blood.

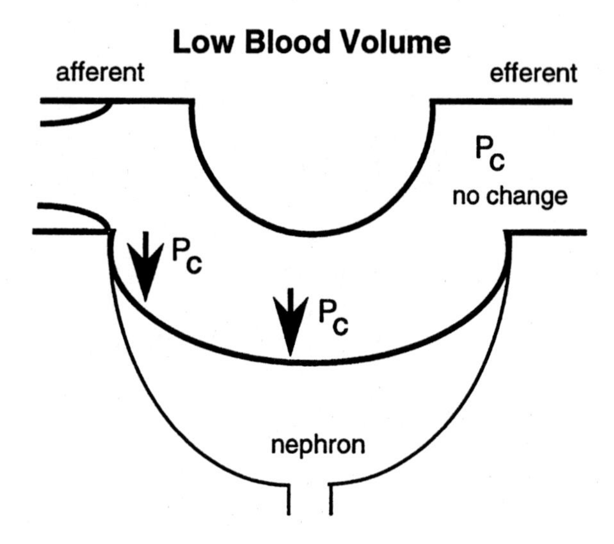

a. What do Steve's kidneys do to adapt to the sudden loss of blood?

3. The Effects of Dehydration

Imagine yourself sitting in a hot classroom and your teacher won't let you get a drink because he or she is in a grumpy mood. The teacher claims that if you get up to get a drink, then the whole class will be distracted. This goes on all day and you are very thirsty.

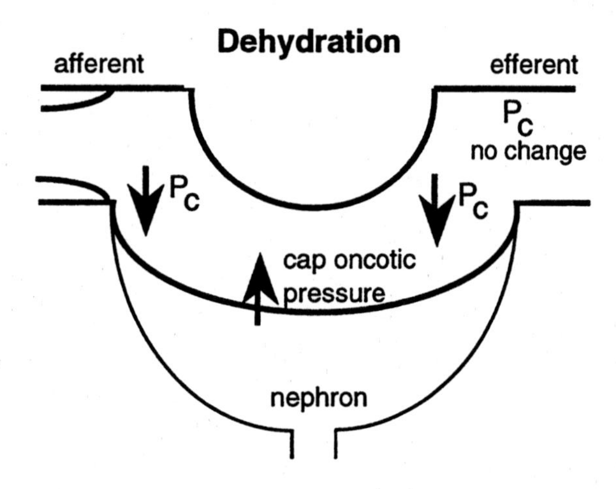

a. How do your kidneys adapt to this situation?

b. What happens to your GFR?

4. The Effects of High Salt Intake

Sally loves salty foods. She tries to hold back and not eat many of her favorites, such as potato chips and popcorn, because she is watching her weight. But when she received a bad grade on her history test, she gave in to her cravings and ate french fries and potato chips.

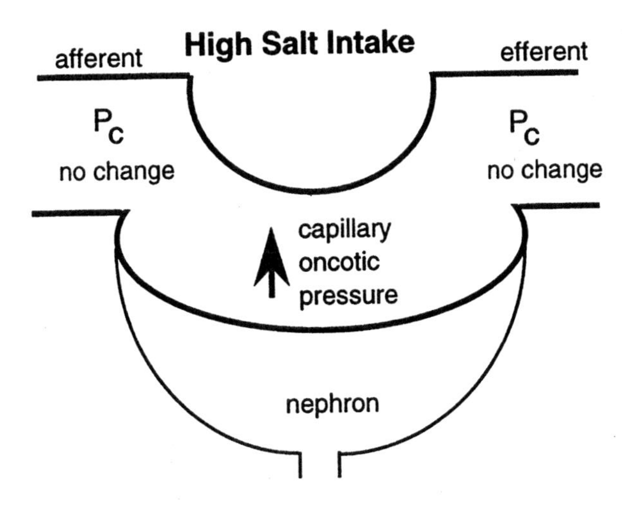

a. Which variable of Starling's equation is most affected?

b. What happens to the plasma flow?

5. Design your own experiment; use the figure below to illustrate the changes in the hydrostatic and oncotic pressures.

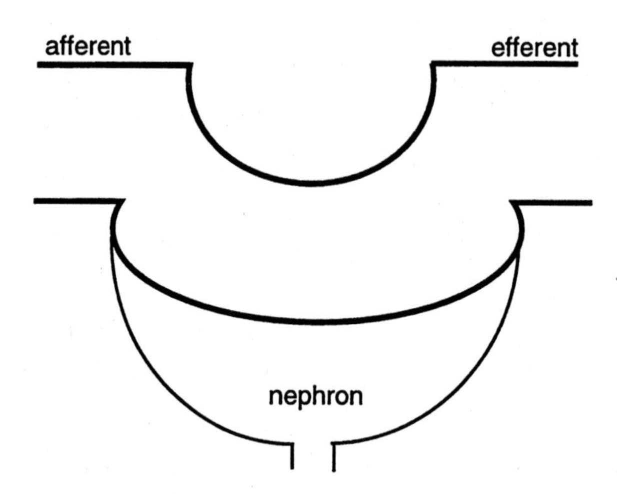

Part II: Applications of Starling's Equation To An Ankle Sprain

This exercise is designed to determine the effects of the standard treatment for an ankle sprain (mobilization, ice, compression, and elevation) on the components of the Starling equation responsible for fluid flow across the capillary membrane.

BACKGROUND

Probably the most universal treatment for an acute injury, such as a sprain, is the application of cold. The treatment also includes mobilization of the joint within the pain-free range of motion, wrapping the injury to apply pressure and elevating the area; the acronym for this procedure is **MICE** or (**M**)obilization, (**I**)ce, (**C**)ompression, and (**E**)levation. The following exercise concentrates on the physiological principles of how these four steps function to improve tissue nutrition based on Starling's equation for fluid flow across the capillary membrane. We now know the steps in treating a sprain; let's look at why these steps are effective.

Inflammation occurs as a result of tissue injury. Inflammation consists of vasodilation of local blood vessels and therefore increased blood flow to the area. Increased capillary permeability occurs, which allows large amounts of pure plasma to leak into the interstitial spaces, culminating in edema. Blood clotting is possible due to leakage of excess amounts of fibrinogen (protein involved in blood clotting) and other proteins leaking out of the capillary. The local inflammatory reaction to a soft-tissue injury also includes an increase in both tissue metabolism and temperature.

The first step is the application of a *cold* pack to the injured area. Many people think that the cold pack acts primarily as an analgesic function (pain relief) because cold does in fact lead to temporary loss of feeling in the local area; it is thought that cold impulses bombard and override the pain receptors. In addition, cold causes vasoconstriction and decreases tissue metabolism. Vasoconstriction occurs to capillary filtration. Cold application is necessary to reduce the metabolic demands of the tissue, restrict hemorrhage, and prevent edema. Essentially, the application of **cold causes afferent arteriole constriction** (the vessel entering the capillary bed is the afferent arteriole). Cold should be applied to an injured area only for 10–15 minutes every half hour to enable the tissue to receive nutrition.

Mobilization of the involved part is beneficial because muscle contraction and movement of the limb within the individual's pain-free range of motion are effective in removing excess fluid, improving the nutritional status of the tissue, and preventing soft-tissue tightness and adhesions. Wrapping the area as a means of **compressing** the injured area increases the interstitial fluid pressure, facilitating the movement of solutes and water from the interstitium into the **lymphatic system.**

The lymphatic system returns protein, water, and electrolytes from the interstitial spaces to the blood. If the lymphatic system has reached its maximum capability for fluid removal, edema will occur. The flow of lymph is dependent on the lymphatic pump and the fluid interstitial pressure. Compression of lymph vessels by muscle contraction, movement of body parts, arterial pulsations, and external compression of the tissue provide a pumping action that enhances lymph flow. Therefore **mobilization and compression increase P_{if}** and are important for the movement of proteins and fluid into the lymphatic system. P_{if} is the pressure in the interstitial fluid space.

Elevation is the final part of the treatment of the injured area. Blood vessels may have been damaged as a result of injury. Elevation of the area assists in returning the blood to the heart by two means, **decreasing the efferent arteriole pressure** and allowing gravity to pull the blood toward the heart.

We have established the function of mobilization, ice, compression, and elevation in treating acute soft-tissue injuries. We can explain the effectiveness of MICE further by looking at fluid movement across the capillary wall as a result of hydrostatic or oncotic pressure differences. The Starling equation describes this relationship:

$$\text{Flow} = k\ [(P_c - P_{if}) - s\,(\Pi_c - \Pi_{if})]$$

Where P_c = capillary hydrostatic pressure
P_{if} = interstitial hydrostatic pressure
Π_c = capillary oncotic pressure
Π_{if} = interstitial oncotic pressure
k = filtration coefficient
s = osmotic reflection coefficient for plasma proteins

Figure 1. Pressures responsible for fluid flow across the capillary membrane

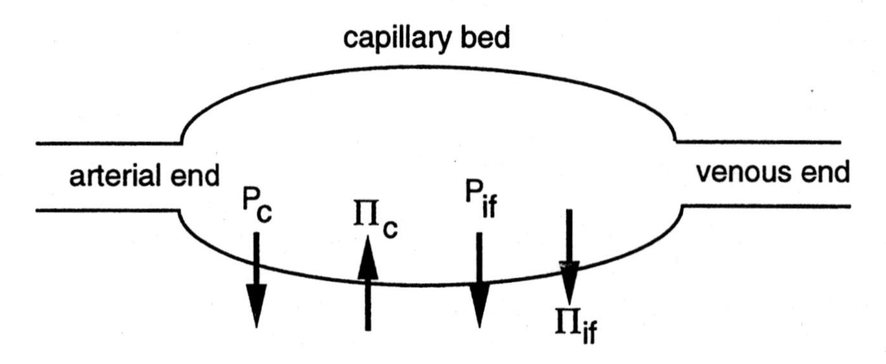

The parameters of the Starling equation are illustrated in Figure 1. The filtration coefficient (k) of capillaries is proportional to the product of the capillary permeability to water and the surface area. The coefficient can be increased by opening capillary beds, thereby increasing the surface area. The value of the osmotic reflection coefficient for plasma proteins (s) ranges between 0 and 1. Zero represents total membrane permeability to proteins, and 1 represents total impermeability to protein. The pressure within the capillary (P_c) forces fluid out of the capillary. The fluid pressure in the interstitium (P_{if}) is negative under normal conditions and therefore pulls fluid out of the capillary. Capillary oncotic pressure (Π_c) develops in the capillary due to the presence of proteins; the capillary oncotic pressure pulls fluid into the capillary. The interstitium also has an oncotic pressure (Π_{if}), which pulls fluid out of the capillary.

The forces counteract one another; under normal conditions, there is a small net flow of fluid out of the capillary. However, if large amounts of fluid leak out of the capillary in excess of the rate at which lymph can carry the fluid back to the blood, edema develops. The rate of lymph flow can be increased by any form of compression (previously mentioned) and factors that will increase interstitial fluid pressure (P_{if}). These factors include increasing capillary pressure (P_c), decreasing capillary oncotic pressure (Π_c), increasing protein in the interstitial fluid (Π_{if}) and increasing the permeability of the capillaries. The following exercise was developed for students to further understand the relationship between the different forces responsible for fluid flow across the capillary membrane.

PROCEDURE

1. Describe the changes in Starling forces (increase/ decrease/ no change) that you would expect to occur during the following situations and record them in Table 2:

   a. Application of *cold* to reduce blood flow to the area, prevent edema, and reduce the metabolic demands of the tissue.

   b. *Mobilization* within the individual's pain free range of motion and *compression* of the area with an elastic bandage. Wrapping the injured area with an elastic bandage or some other wrap applies *pressure* to the lymph vessels and, as does *mobilization*, facilitates the movement of lymph.

   c. *Elevation* of the involved area increases blood flow back to the heart.

Table 2. Changes in the parameters of the Starling equation with MICE treatment (increase, decrease, or no change)

Treatment	P_{if}	P_c	Π_{if}	Π_c	k
Cold					
Mobilization					
Compression					
Elevation					

2. Explain the hydrostatic and osmotic forces responsible for fluid flow across the capillary membrane. Draw a picture representing the forces causes flow in and out of the capillary. Why can P_{if} be considered a force that causes flow out of the capillary?

3. If the net movement of fluid is out of the capillary, why doesn't edema develop?

4. Explain the use of compression and mobilization in terms of P_{if}

ACKNOWLEDGMENTS

We would like to thank Linda I. Anderson for her assistance with the illustrations and David W. Rodenbaugh and Cynthia M. Odenweller for their careful and thorough editing.